Topics in
Current Physics

27

Topics in Current Physics   Founded by Helmut K. V. Lotsch

1 **Beam-Foil Spectroscopy**
Editor: S. Bashkin

2 **Modern Three-Hadron Physics**
Editor: A. W. Thomas

3 **Dynamics of Solids and Liquids by Neutron Scattering**
Editors: S. W. Lovesey and T. Springer

4 **Electron Spectroscopy for Surface Analysis**
Editor: H. Ibach

5 **Structure and Collisions of Ions and Atoms**
Editor: I. A. Sellin

6 **Neutron Diffraction**
Editor: H. Dachs

7 **Monte Carlo Methods** in Statistical Physics
Editor: K. Binder

8 **Ocean Acoustics**
Editor: J. A. DeSanto

9 **Inverse Source Problems** in Optics
Editor: H. P. Baltes

10 **Synchrotron Radiation**
Techniques and Applications
Editor: C. Kunz

11 **Raman Spectroscopy** of Gases and Liquids
Editor: A. Weber

12 **Positrons in Solids**
Editor: P. Hautojärvi

13 **Computer Processing of Electron Microscope Images**
Editor: P. W. Hawkes

14 **Excitons** Editor: K. Cho

15 Physics of **Superionic Conductors**
Editor: M. B. Salamon

16 **Aerosol Microphysics I**
Particle Interactions
Editor: W. H. Marlow

17 **Solitons**
Editors: R. Bullough, P. Caudrey

18 **Properties of Magnetic Electron Lenses**
Editor: P. W. Hawkes

19 Theory of **Chemisorption**
Editor: J. R. Smith

20 **Inverse Scattering Problems** in Optics
Editor: H. P. Baltes

21 **Coherent Nonlinear Optics**
Recent Advances
Editors: M. S. Feld and V. S. Letokhov

22 **Electromagnetic Theory of Gratings**
Editor: R. Petit

23 **Structural Phase Transitions I**
Editors: K. A. Müller and H. Thomas

24 **Amorphous Solids**
Low-Temperature Properties
Editor: W. A. Phillips

25 **Mössbauer Spectroscopy II**
The Exotic Side of the Method
Editor: U. Gonser

26 **Crystal Cohesion and Conformational Energies** Editor: R. M. Metzger

27 **Dissipative Systems in Quantum Optics**
Resonance Fluorescence, Optical Bistability, Superfluorescence
Editor: R. Bonifacio

28 **The Stratospheric Aerosol Layer**
Editor: R. C. Whitten

29 **Aerosol Microphysics II**
Chemical Physics of Microparticles
Editor: W. H. Marlow

30 **Real Space Renormalization**
Editors: J. M. van Leeuwen and T. W. Burkhardt

31 **Hyperfine Interactions of Radioactive Nuclei**
Editor: J. Christiansen

32 **Superconductivity in Ternary Compounds I**
Structural, Electronic, and Lattice Properties
Editors: B. Maple and O. Fischer

Dissipative Systems in Quantum Optics

Resonance Fluorescence, Optical Bistability, Superfluorescence

Edited by R. Bonifacio

With Contributions by
R. Bonifacio J. D. Cresser H. M. Gibbs J. Häger
G. Leuchs L. A. Lugiato S. L. McCall B. R. Mollow
M. Rateike Q. H. F. Vrehen H. Walther

With 60 Figures

Springer-Verlag Berlin Heidelberg New York 1982

Professor Rodolfo Bonifacio

Universita degli Studi de Milano, Istituto Scienze Fisiche "Aldo Pontremoli", Via Celoria 16
I-20133 Milano, Italy

ISBN 3-540-11062-3 Springer-Verlag Berlin Heidelberg New York
ISBN 0-387-11062-3 Springer-Verlag New York Heidelberg Berlin

Library of Congress Cataloging in Publication Data. Main entry under title: Dissipative systems in quantum optics. (Topics in current physics ; 27) Bibliography: p. Includes index. 1. Quantum optics. 2. Optical bistability. 3. Fluorescence. I. Bonifacio, R. II. Lugiato, L. A. (Luigi A.), 1944–. III. Series.
QC446.2.D57 535'.35 81-14570 AACR2

Offset printing and bookbinding: Konrad Triltsch, Graphischer Betrieb, Würzburg.
2153/3130-543210

Preface

In studying the radiation-matter interaction, one can take two different approaches. The first is typical of spectroscopy: one considers the interaction between radiation and a *single* atom, i.e., one studies those phenomena in which the presence of other atoms is irrelevant. The other attitude consists, in contrast, in studying those phenomena which arise just from the simultaneous presence of *many atoms*. In fact, all the atoms interact with the same electromagnetic field; under suitable conditions, this situation creates strong atom-atom correlations, which in turn give rise to a cooperative behavior of the system as a whole. Cooperative means that the overall behavior is quite different from the superposition of the effects arising from single atoms and is completely unpredictable if one neglects the coupling between the atoms induced by their common electromagnetic field.

This book contains five complete and up-to-date contributions on the theory and experiments of three coherence effects in radiation-matter interaction: resonance fluorescences, optical bistability, and superfluorescence. They have raised increasing interest in recent years from both a fundamental and an applicative viewpoint. Even if their phenomenology appears completely different, these effects belong in the same book because they are striking examples of *open systems* driven far from thermal equilibrium, as those considered in Haken's *synergetics* and in Prigogine's theory of *dissipative structures*. This aspect is discussed in the introducting chapter, in which we outline the basic physics and the essential features which unify these three effects.

The body of the book is composed of two theoretical papers and three experimental ones written by well-known experts who describe the state of the art in the three subjects. The experimental articles are self-contained since they contain enough theory to permit understanding the meaning of the experimental results, whose applications range from laser spectroscopy to optical memory elements.

The aim of all the chapters is to give a complete account of the subjects in a style understandable to graduate students in physics, and to stimulate further theoretical and experimental research.

Milano, October 1981

Rodolfo Bonifacio

Contents

List of Contributors

Bonifacio, Rodolfo

Università degli Studi di Milano, Istituto die Scienze Fisiche "Aldo Pontremoli", Via Celoria 16, I-20133 Milano, Italy

Cresser, James Donald

Max-Planck-Institut für Quantenoptik, D-8046 Garching, Fed. Rep. of Germany

Gibbs, Hyatt McDonald

Optical Sciences Center, University of Arizona, Tucson, AZ 85721, USA

Häger, Jürgen

Max-Planck-Institut für Quantenoptik, D-8046 Garching, Fed. Rep. of Germany

Leuchs, Gerd

Universität München, Sektion Physik, Am Coulombwall 1, D-8046 Garching, Fed. Rep. of Germany

Lugiato, Luigi A.

Università degli Studi di Milano, Istituto di Scienze Fisiche "Aldo Pontremoli", Via Celoria 16, I-20133 Milano, Italy

McCall, Samuel L.

Bell Laboratories, Murray Hill, NJ 07974, USA

Mollow, Benjamin R.

The University of Massachusetts at Boston, Department of Physics, Boston, MA 02125, USA

Rateike, Matthias

Universität München, Sektion Physik, Am Coulombwall 1, D-8046 Garching, Fed. Rep. of Germany

Vrehen, Quirin H.F.

Philips Research Laboratories, Eindhoven, The Netherlands

Walther, Herbert

Universität München, Sektion Physik, Am Coulombwall 1, and Max-Planck-Institut für Quantenoptik, D-8046 Garching, Fed. Rep. of Germany

1. Introduction: What are Resonance Fluorescence, Optical Bistability, and Superfluorescence

R. Bonifacio and L. A. Lugiato

1.1 General Remarks

Radiation-matter interaction is a topic in synergetics [1.1] as well as in the theory of dissipative structures [1.2]. In fact, the atoms plus the electromagnetic field form an *open* system which exhibits phenomena analogous to phase transitions but far from thermodynamic equilibrium. From the mathematical viewpoint these situations exhibiting cooperative behavior are typically described by *nonlinear* equations with suitable boundary conditions, which can be essential for the rise of cooperative behavior, as in the laser or in optical bistability.

Let λ be a parameter that measures the strength of the interaction of an open system with the external world. When λ is small, the effect of nonlinearity is also small. The system is in a "quasi-equilibrium" situation in which it shows a linear dependence on λ. This behavior can be considered as the direct "analytic continuation" of the thermodynamic equilibrium state. On the other hand, when λ increases, the system becomes more and more unbalanced. In correspondence to a suitable threshold value λ_c, the quasi-equilibrium state becomes unstable, and the system can show, roughly speaking, three different types of behavior.

a) It can perform a transition to a new kind of steady state that "bifurcates" from the quasi-equilibrium state. In this case, the behavior is similar to first- or second-order phase transitions in an equilibrium system, and λ_c can be characterized as a critical value of λ. Of course, the fluctuations play a major role in the transition at the critical point.

b) It can approach a nonstationary situation, that is, one which is periodic in time. In other words, the state of the system in the long time limit is not represented by a fixed point in the phase space (steady state), but by a limit cycle. In this case the system shows a pulsing behavior, or more precisely a "self-pulsing" behavior, because it does not arise from external manipulation but is spontaneously generated by the self-organization of the system itself.

c) It can approach nonperiodic behavior that does not exhibit any kind of regularity in time, so that it is called chaotic.

These three types of transitions can also appear in succession when λ is increased, as a result of successive bifurcations.

Let us mention a few examples of cooperative phenomena in quantum optics, which we shall classify according to the process (stimulated emission, spontaneous emission, absorption) which dominates in each phenomenon.

a) The typical example for stimulated emission is the *laser*, which is also the best studied in all its variants.

b) Spontaneous emission from a collection of N atoms can be cooperative, and in this case one has *superradiance* or *superfluorescence*.

c) Quite recently it has been discovered that absorption can also give rise to relevant cooperative effects, as *optical bistability*.

In all these phenomena, the competition between cooperative behavior and one-atom behavior is described by the parameter C, which is the ratio between the cooperative decay rate $\gamma_R = \tau_R^{-1}$ of pure superfluorescence [1.3] and the decay rate $\gamma_\perp$ of the single atom (i.e., $2\gamma_\perp$ is the atomic linewidth)

$$C = \gamma_R/2\gamma_\perp \quad . \tag{1.1}$$

The parameter C is proportional to the atomic density ρ, as one sees from the alternative expression

$$C = \alpha_{abs}L/2T \tag{1.2}$$

where α_{abs} is the linear absorption coefficient (which is proportional to ρ), L is the length of the atomic sample, and T is the transmissivity coefficient of the mirrors. For $C \ll 1$ the atoms evolve independently of one another, whereas for $C \gg 1$ cooperative behavior is dominant.

This book does not consider the laser, that has already been extensively treated in several outstanding monographs (see, e.g., [1.4,5]). It deals with three different topics, resonance fluorescence, optical bistability, and superfluorescence, that are treated from both a theoretical and an experimental viewpoint. These are three examples of open systems driven out of thermal equilibrium. In this Introduction we will outline the essential features of these phenomena. From this description the reason will emerge clearly why these different topics have been collected in the same volume. The historical elements that we shall give here are less than essential. For an adequate description of the history of each topic, refer to the individual chapters.

1.2 Resonance Fluorescence

Let us consider an atom driven by a coherent electromagnetic (em) field resonant or quasi-resonant with the atomic frequency. According to *classical* electrodynamics, the fluorescent light diffused by the atom has the same frequency as the incident field. Surprisingly enough, a complete theoretical description of the light diffused by a *quantum-mechanical* two-level atom under the action of a coherent field was first given relatively recently. In fact, even if there had been earlier contributions, a satisfactory description was achieved by MOLLOW in 1969 [1.6]. The resulting picture is as follows. For a small enough incident field the behavior coincides with that of classical electrodynamics, i.e., the scattering of photons by the atoms is perfectly elastic. Increasing the incident field, an inelastic component begins to appear. It consists of a Lorentzian peak, centered on the driving field frequency, having a width equal to the natural linewidth γ. This is not surprising, because the line shape is also a Lorentzian of width γ in the case of normal blackbody radiation (here, however, the Lorentzian is centered on the atomic transition frequency and is there obviously no elastic component). The interesting behavior occurs when the incident field is so strong, that the so-called Rabi frequency of the incident field exceeds the natural linewidth γ. The Rabi frequency is defined as

$$\Omega = \frac{\mu E}{\hbar} \quad , \tag{1.3}$$

where μ is the modulus of the atomic dipole moment of the two-level atom and E is the incident field amplitude. For simplicity, let us assume that the incident field frequency perfectly coincides with the atomic transition frequency. When $\Omega > \gamma$ the spectrum becomes three peaked, i.e., the central peak is accompanied by two symmetrical Lorentzian sidebands of width $(3/2)\gamma$. The shift between the central frequency and the sidebands is equal to the Rabi frequency Ω. The appearance of a three-peaked spectrum when the incident field intensity is increased enough is called the *dynamical stark effect*. Under these conditions of an intense incident field, the elastic component is negligible with respect to the three-peaked inelastic component.

Clearly, resonance fluorescence is not a cooperative phenomenon because it is completely described in terms of a single atom. However, it is the simplest example of a quantum-mechanical open system driven by an external field. When this field is weak, the system is in the "quasi-thermodynamic" branch and the incoherent part of the spectrum is a single Lorentzian of width γ as in thermal blackbody radiation. On the other hand, when the incident field is increased, the system is driven further and further out of equilibrium. In fact, the probability of occupation of the upper state at steady state, which is zero in absence of external field, increases with the incident field and becomes equal to the probability of the lower state for $\Omega \gg \gamma$ (infinite saturation limit). For $\Omega = \gamma$ we have a transition from a one-peaked to a three-peaked spectrum, which appears when the Rabi frequency becomes larger than the natural linewidth so that the sidebands emerge from the central Lorentzian. Hence in this case we also have the appearance of a new structure when the system is driven far enough from thermal equilibrium. In this phenomenon the transition is smooth like in second-order phase transitions. The first observation of the three-peaked spectrum was made by SCHUDA, STROUD and HERCHER in Rochester in 1974 [1.7]. More precise and systematic data were later produced by a group around WALTHER in Garching and by a group around EZEKIEL at MIT.

1.3 Superfluorescence

In ordinary light sources the emitted radiation intensity is simply proportional to the number N of atoms because the phases of the atomic dipoles are completely random. On the other hand, an ordered array of atomic dipoles emits *coherently*, so that the intensity is proportional to N^2. Such an array can be easily obtained by irradiating the atomic sample with a coherent light pulse (*coherent excitation*). However, emission as N^2 is not interesting per se because usually it is a trivial consequence of the preparation of the system. In phenomena such as free induction decay the radiation is proportional to N^2, but the time evolution of the atomic system crudely consists of the progressive loss of the initial phase coherence, i.e., of the dephasing of the atomic dipoles. Only under special conditions the initial coherent preparation gives rise to a cooperative behavior of the atomic system which produces an abnormally fast emission of radiation. In this case one has *superradiance* (SR) predicted by DICKE in 1954 [1.8]. The emission occurs in a time proportional to N^{-1} and is strongly directional, the radiation being assembled around the direction of the exciting coherent pulse.

In this fundamental paper DICKE also predicted the possibility that an *incoherently excited* atomic system (i.e., an atomic sample prepared with population inversion but without macroscopic polarization) emits cooperative spontaneous emission *provided that* the atoms are confined in a volume smaller than a cubic wavelength. This condition has practically no macroscopic meaning at infrared or optical wavelengths. Hence the question arose whether the same phenomenon can occur in an extended sample or not. A positive answer was first given in [1.3] where the specific conditions for the observation of cooperative spontaneous emission from an incoherently excited, pencil-shaped extended system have been explicitly stated. These conditions have been fulfilled in the experiments in Cs described by VREHEN and GIBBS in their contribution in this book. We invented the name of *superfluorescence* for this phenomenon in order to distinguish it from superradiance. In fact, in the case of superfluorescence the atomic dipoles are initially random, so that the system begins to radiate by *normal fluorescence* proportional to N. However, the em field creates strong atom-atom correlations which induce cooperative emission of the same type of superradiance, that occurs, however, in both the positive and the negative directions of the longitudinal axis. Roughly speaking, the weak field emitted by normal fluorescence produces a weak polarization in the medium, which increases the field, which in turn increases the polarization, and so on. This runaway behavior finally produces a macroscopic polarization which gives rise to cooperative emission. Hence in superfluorescence the atomic dipoles put themselves in order. This is completely different from superradiance, in which the system has a macroscopic polarization just by preparation, and the consequent emission is a purely *classical* dipole radiation. In contrast, in superfluorescence the macroscopic dipole and therefore the coherent cooperative emission is spontaneously created by the self-organization of the system itself, because the radiation is started by *quantum-mechanical* incoherent spontaneous emission. Thus superfluorescence is a unique example of transient phenomenon in radiation-matter interaction, in which the macroscopic system exhibits self-organization.

Let us now specify the superfluorescence conditions, as stated in [1.3] for a pencil-shaped sample with Fresnel number of order unity. These conditions involve the cooperative decay rate γ_R which, apart from a numerical factor of order unity, is defined as the product of the purely radiative linewidth γ times the number of atoms, times a geometrical factor given by the ratio of the diffraction solid angle λ_0^2/S over the total solid angle 4π,

$$\gamma_R = \frac{3}{2} N\gamma(\lambda_0^2/4\pi S)$$

$$= \frac{3}{8} \gamma\rho L\lambda_0^2 \quad , \tag{1.4}$$

where λ_0 is the wavelength of the radiation, L and S are the length and the sectional area of the atomic sample, respectively, and ρ is the atomic density. Note that γ_R is proportional to N as well as to ρ.

Now the superfluorescence conditions are

$$\gamma_\perp \ll \gamma_R \lesssim \frac{c}{L} \tag{1.5}$$

where $\gamma_\perp = T_2^{-1}$ is the homogeneous linewidth and L/c is the transit time of the photons in the sample. Note that (1.5) implies in particular $T_2 \gg L/c$, contrary to what occurs in the usual laser amplifiers. The left-hand part of (1.5) amounts to the condition $C \gg 1$, as one sees from (1.1), or equivalently to the *high-gain condition*

$$\alpha L \gg 1 \tag{1.6}$$

where α is the gain per unit length

$$\alpha = \frac{3}{8} \frac{\gamma}{\gamma_\perp} \rho\lambda_0^2 \quad , \tag{1.7}$$

which in amplifying systems is the courterpart of the absorption coefficient α_{abs} [cf. (1.2) in which T = 1 because there are no mirrors].

The condition $C \gg 1$ ensures that cooperative emission dominates over incoherent one-atom emission. For $C < 1$, the atoms emit independently of one another and the emission law is purely exponential. This means that the system is in the thermodynamic branch and the emission is blackbody radiation. For $C > 1$ the system is initially far from thermodynamic equilibrium. When C is increased beyond the critical threshold $C = 1$, the system gradually performs a transition to a stage in which the emission has a completely different character. In fact, it consists of a single pulse or of a sequence of a few pulses. The height of the first pulse is proportional to N^2, and its width is of the order of a few τ_R, where $\tau_R = \gamma_R^{-1} \propto N^{-1}$. Using (1.4) the right-hand part of condition (1.5) can be rephased as it follows

$$L \lesssim L_c \quad , \tag{1.8}$$

$$L_c = \left(\frac{8}{3} \frac{c}{\rho \gamma \lambda_0^2} \right) \quad ,$$

where L_c is the cooperation length introduced by ARECCHI and COURTENS [1.9]. Hence the atomic sample must be *shorter than the cooperation length*. By combining (1.6) and (1.8) one finds the necessary condition

$$\alpha L_c \gg 1 \quad , \tag{1.9}$$

i.e., the gain in cooperation length must be large. Single-pulse emission occurs for $L \ll L_c$, in which case the shape of the pulse is similar to that of a squared hyperbolic secant. We have called this single-pulse emission *pure superfluorescence*. For $L \sim L_c$ the superfluorescent emission consists of few pulses (ringing). When the system is inhomogeneously broadened, in the left-hand part of (1.5) $\gamma_\perp$ must be replaced by the inhomogeneous linewidth $(T_2^*)^{-1}$. The first observation of superfluorescence was performed by SKRIBANOWITZ et al. [1.10] in 1973 at MIT. In this experiment the emission showed ringing and the inhomogeneous decay time was comparable to the emission time. The first observation of pure superfluorescence was obtained in experiments by VREHEN and GIBBS [1.11] in 1976-1977 at Philips Laboratories in Eindhoven, in which the conditions described above were fully satisfied for the first time.

A satisfactory theory of superfluorescence to our opinion does not yet exist, mainly for two reasons. First, a complete theory of this phenomenon must be necessarily quantum mechanical in order to properly describe the transition from the initial isotropic normal fluorescent stage to the following directional superfluorescent cooperative emission stage. Second, as clearly indicated by the experimental data of VREHEN and GIBBS, despite the directional character of the emission superfluorescence must be described in full three-dimensional terms. In fact, the transverse effects pointed out by these authors prove that the transverse profile of the field inside the active volume plays an essential role in determining the details of the emission. This is the reason why the experimental data show only a qualitative agreement with existing theories, which at most correctly predict either the ringing behavior or the delay time, but never both simultaneously. In order to do that from first principles one needs a quantum-mechanical theory with transverse effects and propagation, which at the moment does not exist to our knowledge. These remarks also explain why this book does not contain a chapter with a detailed description of the available theories of superfluorescence. Some discussion of theoretical results is contained in the Chap.6.

1.4 Optical Bistability

Let us consider a coherent cw laser beam that is injected into an optical cavity, resonant or quasi-resonant with the incident light. As is well known, when the cavity is empty, the transmitted intensity I_T is proportional to the incident intensity I_I, and in particular $I_T = I_I$ for exact resonance. The interesting situation occurs, of course, when the cavity is filled with atomic material resonant or quasi-resonant with the incident field. In this situation I_T becomes a nonlinear function of I_I. In particular when the cooperativity parameter C exceeds a suitable critical value which depends on the atomic and cavity detunings and on the inhomogeneous linewidth, I_T varies discontinuously under adiabatic variation of I_I, showing a hysteresis cycle. In this cycle one can distinguish a low-transmission and a high-transmission branch, which means a bistable situation. For small incident field the system is in the low-transmission branch, and the response of the system is linear with respect to the incident intensity. This is the "quasi-equilibrium situation". On the other hand, when I_I is large enough the system exhibits a discontinuous transition by switching to the high-transmission branch. Thus, the resulting picture is closely similar to first-order phase transitions in equilibrium systems. This behavior arises from the combined action of the nonlinearity of the atomic medium and of the feedback action arising from the mirrors of the cavity. One usually distinguishes between the situation in which dispersion is negligible (purely absorptive optical bistability) and the converse situation in which absorption is negligible (purely dispersive optical bistability). The switching mechanism is different in the two cases. In fact, in absorptive optical bistability it occurs when the field internal to the cavity becomes strong enough to bleach the absorber; as we said, this bleaching does not arise gradually, but abruptly. On the other hand, in the dispersive case the switching occurs when the change in cavity frequency induced by the dispersive part of the atomic polarization is such that the cavity attains resonance with the incident field, thereby producing transparency.

When there is switching, the spectrum of the transmitted light also undergoes a discontinuous transition. Under suitable conditions and when the incident field is increased from zero to beyond the switching threshold, the behavior of the spectrum of transmitted light shows some analogies with the behavior of the spectrum of single-atom resonance fluorescence. However, it also shows many striking differences that arise from the cooperative nature of optical bistability, namely, crossing the switching threshold the incoherent part of the spectrum of transmitted light abruptly changes from a single narrow line to a triplet with well-separated sidebands (discontinuous dynamical stark effect). The phenomenon of optical bistability was first predicted in 1969 by SZÖKE and collaborators [1.12] and was first observed by GIBBS, McCALL, and VENKATESAN in 1976 at Bell Laboratories [1.13]. The interest that this subject has raised in the recent years is both of technological and of theoretical character. The technological interest arises from the fact that this system presents itself as a natural candidate to work as an optical memory, optical transistor, and so on. In fact, the most recent efforts in this direction have been devoted to the construction of miniaturized, fast-switching optical bistable devices. On the theoretical side, optical bistability has renewed the interest that fifteen years ago was raised by the laser, by offering to the community of theoreticians working in quantum optics a novel cooperative phenomenon to be analyzed in all its details.

More recently, in collaboration with GRONCHI we predicted that under suitable conditions a part of the high-transmission branch becomes unstable [1.14]. In this case, the system approaches a limit cycle behavior, in which the transmitted light is no longer stationary but consists of a periodic sequence of short pulses. Again, this phenomenon is interesting both from a theoretical and from a practical viewpoint. The theoretical interest arises from the fact that it provides a striking example of the self-pulsing behavior that we mentioned in Sect.1.1. On the other hand, from a practical viewpoint this phenomenon suggests a device to convert cw light into pulsed light.

The origin of this self-pulsing behavior lies in the fact that under suitable conditions some of the off-resonance frequencies of the cavity become unstable and therefore experience gain. Thus, part of the incident energy gets transferred from the resonant frequency to the unstable frequencies which are amplified. Hence, under these conditions the system works as a novel type of laser without population inversion. IKEDA [1.15] has also predicted that under suitable conditions a multistable optical system in a ring cavity shows chaotic behavior.

References

1.1 H. Haken: *Synergetics. An Introduction*, 2nd ed., Springer Series in Synergetics, Vol.1 (Springer, Berlin, Heidelberg, New York 1978)
1.2 G. Nicolis, I. Prigogine: *Self-Organization in Non-Equilibrium Systems* (Wiley, New York 1977)
1.3 R. Bonifacio, P. Schwendimann: Lett. Nuovo Cimento *3*, 509, 512 (1970); R. Bonifacio, P. Schwendimann, F. Haake: Phys. Rev. A*4*, 382, 854 (1971); R. Bonifacio, L.A. Lugiato: Phys. Rev. A*11*, 1507 (1975); *12*, 587 (1975)
1.4 H. Haken: "Laser Theory", in *Light and Matter Ic*, ed. by L. Genzel, Handbuch der Physik, Vol.XXV/2c (Springer, Berlin, Heidelberg, New York 1970)
1.5 M. Sargent, M.O. Scully, W.E. Lamb, Jr.: *Laser Physics* (Addison-Wesley, Reading, MA 1974)
1.6 B.R. Mollow: Phys. Rev. *188*, 1969 (1969)
1.7 F. Schuda, C.R. Stroud, Jr., M. Hercher: J. Phys. B*7*, L 198 (1974)
1.8 R.H. Dicke: Phys. Rev. *93*, 99 (1954)
1.9 F.T. Arecchi, E. Courtens: Phys. Rev. A*2*, 1730 (1970)
1.10 N. Skribanawitz, I.P. Herman, J:C. MacGillivray, M.S. Feld: Phys. Rev. Lett. *30*, 309 (1973)
1.11 H.M. Gibbs, Q.H.F. Vrehen, H.M.J. Hikspoors: Phys. Rev. Lett. *39*, 547 (1977)
1.12 A. Szöke, V. Daneu, J. Goldhar, N.A. Kurnit: Appl. Phys. Lett. *15*, 376 (1969)
1.13 H.M. Gibbs, S.L. McCall, T.N.C. Venkatesan: Phys. Rev. Lett. *36*, 113 (1976)
1.14 R. Bonifacio, M. Gronchi, L.A. Lugiato: Opt. Commun. *30*, 129 (1979)
1.15 K. Ikeda: Opt. Commun. *30*, 257 (1979)

2. Intensity-Dependent Resonance Light Scattering

B. R. Mollow

Interesting modifications of the spectrum of (near-) resonance scattered light occur when the incident field intensity is large enough to induce atomic transitions at a rate comparable to or greater than the rate of relaxation-induced transitions, i.e., when the absorption line shape of the incident field is broadened by its own intensity. Treated here will be the case of a single, stationary, homogeneously broadened atom, driven by an incident field which resonantly couples only two atomic states $|0\rangle$ and $|1\rangle$, though coupling to other levels through relaxation processes will be allowed.

2.1 General Method of Solution

2.1.1 Approximations and Limiting Assumptions

The incident field $E(t) = E \cos\omega t$ will be assumed to be fully monochromatic and coherent, with perfect amplitude and phase stability. Its frequency ω will be assumed to be near enough to the atomic resonance frequency $\omega_{10} = (E_1 - E_0)/\hbar$ to justify the rotating wave approximation, i.e., $\Delta \equiv \omega - \omega_{10} \ll \omega$. The field intensity will be allowed to range from zero to a value which fully saturates the transition, but not so high as to produce harmonics in the emission spectrum. The conditions on the field frequency ω and amplitude E are equivalent to the statement that the Rabi frequency

$$\Omega' = (\Omega^2 + \Delta^2)^{\frac{1}{2}} \quad , \tag{2.1}$$

where Ω is the power-broadening parameter

$$\Omega = E\mu_{10}/\hbar \tag{2.2}$$

(with μ_{10} the electric dipole matrix element connecting the states $|1\rangle$ and $|10\rangle$) may have any value relative to the homogeneous width of the transition, but must be small compared to the (optical) field frequency, i.e., $\Omega' \ll \omega$.

In the case of purely radiative damping, no further conditions need be imposed. For collisional and other relaxation mechanisms, on the other hand, it must be required that the mean collision rate (or other relaxation-induced width) be small compared to ω, and furthermore that the *duration* of the collision or other damping event (as distinct from the time interval *between* collisions, which is simply the reciprocal of the width) be small compared to Ω'. The latter condition is necessary to justify the *impact approximation*, in which the collision is treated as a sudden event. For purely radiative damping, the analogous conditions are invariably satisfied as long as the assumptions of the preceding paragraph hold; for the natural width of optical emission lines is invariably small compared to ω, while the duration of the damping event — that of photon emission — must be taken as the optical period ω^{-1}.

2.1.2 Optical Bloch Equations

The approximations and assumptions described above are precisely the ones needed to justify the *optical Bloch equations* for the atomic density matrix, in which the atomic relaxation is described by simple damping parameters. In the absence of an incident field, these equations have the general form

$$(d/dt + i\omega_{jk} + \kappa'_{jk})\rho_{jk} = 0 \qquad (j \neq k)$$

$$(d/dt + \kappa_j)\rho_{jj} = \sum_k \rho_{kk}\kappa_{kj} \quad , \tag{2.3}$$

where $\omega_{jk} \equiv (E_j - E_k)/\hbar$, $\kappa'_{jk} = \kappa'_{kj}$, and $\kappa_j = \sum_k \kappa_{jk}$. In addition to the radiative case [for which $\kappa'_{jk} = (\kappa_j + \kappa_k)/2$] and the collisional case, it should be noted that the action of an incoherent broadband pumping field may be represented by relaxation constants.

The four denstiy matrix elements which refer to the single pair of states $|1>$ and $|0>$ which are resonantly coupled by the incident field $E \cos\omega t$ obey the equations, in the rotating wave approximation,

$$(d/dt + i\omega_{10} + \kappa'_{10})\rho_{10} = i\Omega e^{-i\omega t}(\rho_{00} - \rho_{11})/2$$

$$(d/dt - i\omega_{10} + \kappa'_{10})\rho_{01} = -i\Omega e^{i\omega t}(\rho_{00} - \rho_{11})/2$$

$$(d/dt + \kappa_1)\rho_{11} = \sum_j \rho_{jj}\kappa_{j1} - i\Omega e^{i\omega t}\rho_{10}/2 + i\Omega e^{-i\omega t}\rho_{01}/2 \tag{2.4}$$

$$(d/dt + \kappa_0)\rho_{00} = \sum_j \rho_{jj}\kappa_{j0} + i\Omega e^{i\omega t}\rho_{10}/2 - i\Omega e^{-i\omega t}\rho_{01}/2 \quad ,$$

where the state summations in general include the states $|0>$ and $|1>$, and all other states of the atom as well.

2.1.3 The Use of a c-Number Incident Field

The representation of the incident field by a c-number function is a consequence of its being described by a *coherent* state [2.1]. The method in question is *fully quantum mechanical*, and is justified by a canonical transformation [Ref.2.2, Sect.2] which relies upon no approximation of any kind, particularly upon none involving the strength of the incident field or the number of quanta in it. The *radiated* (scattered) field is represented explicitly quantum mechanically, and its action back upon the atom is exactly what gives rise to the radiative damping terms in the optical Bloch equations. Contrary to appearances, the decay of the incident field (more precisely, its attenuation in the forward direction in the traveling wave case under discussion) is fully present, and occurs as the result of destructive interference between the wave functions of the emitted photons and the (unattenuated) c-number field when the aforementioned canonical transformation is inverted.

The use of an n-photon state for the incident field—where n is finite—would be an exact procedure only in the artificial case in which a single isolated atom were placed in a resonant cavity with exactly n photons initially present in a standing-wave mode. By taking the limit $n \to \infty$, however, one does obtain from the n-photon model correct solutions to the physical problem under discussion, in which an atom outside a laser cavity is irradiated by a coherent CW traveling wave emanating from it. It is important to understand that in this realistic physical case, the limit $n \to \infty$ is necessary in order to obtain correct solutions from the n-photon model. The omitted terms of order 1/n (where n is equal, e.g., to the mean number of photons in the cavity of the irradiating laser) are not small corrections but rather are spurious model-dependent artifacts. The c-number method, by contrast, produces no such artifacts and is quite exact for incident fields of arbitrary intensity.

2.1.4 Spectrum of Scattered Field

The spectrum of the field emitted during the atomic transition $|j\rangle \to |k\rangle$ may be evaluated by introducing the transition operators $a_{jk} \equiv |k\rangle\langle j|$, the expectation values of which are just the density matrix elements ρ_{jk}. The steady-state emission spectral density for the transition in question may then be expressed as [2.3]

$$\tilde{g}(\nu) = \int \mathrm{tr}\left\{\bar{\rho} a_{jk}^{\dagger} a_{jk}(t)\right\} e^{i\nu t} dt \quad , \tag{2.5}$$

where $\bar{\rho}$ is the steady-state atomic density operator.

The atomic correlation function in the integrand in (2.5) may be evaluated by means of the quantum *fluctuation-regression theorem* [2.4], which shows that the quantities $R_{jk}(t) \equiv \mathrm{tr}\left\{\bar{\rho} a_{nm}^{\dagger} a_{jk}(t)\right\}$ (with n and m fixed) obey the same equations (2.3,4) as do the density matrix elements $\rho_{jk}(t)$, but with the initial conditions $R_{jk}(0) = \delta_{mk}\bar{\rho}_{jn}$. Thus in the general solution the steady-state density matrix elements appear as constant coefficients; in some cases it may be necessary to treat these as adjustable parameters or to measure them experimentally.

The only nonvanishing density matrix elements in steady state, under the specified conditions, are the populations $\bar{\rho}_{jj}$ and the single off-diagonal matrix element $\bar{\rho}_{10}(t) = \bar{\rho}_{10}\exp(-i\omega t)$ (and its complex conjugate $\bar{\rho}_{01}$). This determines the *coherent part* of the spectrum [2.3],

$$\tilde{g}^{coh}(\nu) = |\bar{\rho}_{10}|^2 2\pi\delta(\nu - \omega) \quad , \tag{2.6}$$

which is present only in the $|1\rangle \rightarrow |0\rangle$ transition spectrum. The remaining part of the spectrum is *incoherent* and, in the case of purely radiative damping, is completely determined by nonlinear terms in the incident field intensity.

2.2 The Closed Two-Level System

When no relaxation-induced transitions (and hence no transitions at all) take place out of the two-dimensional subspace of laser-coupled states $|1\rangle$ - $|0\rangle$, it is possible to obtain closed solutions for the spectrum in relatively simple form. If $\kappa_{01} = 0$, so that no energy-increasing relaxation-induced transitions take place, then the only nonvanishing damping constants in (2.4) are $\kappa_1 = \kappa_{10} \equiv \kappa$ and $\kappa'_{10} = \kappa'_{01} \equiv \kappa'$.

2.2.1 Radiative Relaxation

Of particular interest because it represents a purely electromagnetic process which is solvable [2.3] in a highly nonlinear regime is the case in which the damping is due entirely to the effect of the radiated (scattered) photons back upon the atom. In this case κ is simply the Einstein A coefficient $|\mu_{10}|^2\omega_{10}^3/3\pi\hbar c^3$ (in rationalized units), and $\kappa' = \kappa/2$ [2.5]. The steady-state solutions to (2.4) are [2.3]

$$\bar{\rho}_{10} = i\Omega(\kappa' + i\Delta)/2(\Omega^2/2 + \Delta^2 + \kappa'^2) \quad ,$$

$$\bar{\rho}_{11} = \Omega^2/4(\Omega^2/2 + \Delta^2 + \kappa'^2) \quad , \tag{2.7}$$

$$\bar{\rho}_{01} = \bar{\rho}^*_{10} \quad , \quad \text{and} \quad \bar{\rho}_{00} = 1 - \bar{\rho}_{11} \quad .$$

The emission spectrum is then given quite generally (i.e., for arbitrary field strengths and detunings, subject only to the restrictions in Sect.2.1.1) by the symmetric function [2.3]

$$\tilde{g}(\nu) = \tilde{g}'(\nu - \omega) \quad ,$$

$$\tilde{g}'(\nu) = |\bar{\rho}_{10}|^2 2\pi\delta(\nu) + \bar{\rho}_{11}\kappa\Omega^2(\nu^2 + \Omega^2/2 + \kappa^2)/|f(i\nu)|^2 \tag{2.8}$$

in which

$$|f(i\nu)|^2 = \nu^2(\nu^2 - \Omega'^2 - 5\kappa'^2)^2$$
$$+ \kappa^2[4\nu^2 - (\Omega'^2 + \Delta^2)/2 - \kappa'^2]^2 \quad .$$

In the limit of weak incident fields ($\Omega \ll \kappa$), the spectrum is well approximated as [2.3]

$$\tilde{g}'(\nu) = \frac{\Omega^2/4}{\Delta^2 + \kappa'^2}\left(2\pi\delta(\nu) + \frac{\kappa\Omega^2}{[(\nu - \Delta)^2 + \kappa'^2][(\nu + \Delta)^2 + \kappa'^2]}\right). \tag{2.9}$$

The small incoherent part in this limit has maxima at $\omega - \Delta \equiv \omega_{10}$ and $\omega + \Delta$. The solution (2.9) can be obtained by formal scattering theory [2.2,6]; the incoherent part is the contribution from the process in which two laser photons are absorbed and two photons with energies which sum to 2ω are emitted.

When the condition $\Omega' \gg \kappa'$ is satisfied, the spectral lines are well separated, and the spectrum has the approximate form [2.3,7]

$$\tilde{g}'(\nu) = |\bar{\rho}_{10}|^2 2\pi\delta(\nu) + \frac{2s_0A_0^{inc}}{\nu^2 + s_0^2} + \frac{2\sigma A_+}{(\nu + \Omega')^2 + \sigma^2} + \frac{2\sigma A_-}{(\nu - \Omega')^2 + \sigma^2} \tag{2.10}$$

where

$$A_0^{inc} = \Omega^6/4\Omega'^2(\Omega^2 + 2\Delta^2)^2$$

$$A_+ = A_- = \Omega^4/8\Omega'^2(\Omega^2 + 2\Delta^2) \tag{2.11}$$

and

$$s_0 = (\kappa\Delta^2 + \kappa'\Omega^2)/\Omega'^2$$

$$\sigma = [\kappa\Omega^2 + \kappa'(\Omega^2 + 2\Delta^2)]/2\Omega'^2 \tag{2.12}$$

(with $\kappa' = \kappa/2$).

In the strong field limit ($\Omega \gg \kappa,\Delta$), the coherent part of the spectrum is inappreciable, and the spectrum is described by a function with three peaks, centered at ω, $\omega - \Omega$, and $\omega + \Omega$. Each sideband has an integrated intensity one-half that of the central term ($A_\pm = A_0/2 = 1/8$), and a width 3/2 that of the central term ($\sigma = 3\kappa'/2$, $s_0 = \kappa'$) [2.3]. The predictions of [2.3] have been amply confirmed by experiment [2.8-11].

It should be noted that derivations of the same results as those found in [2.3] have been made with the "dressed atom" approach [2.12,13], which provides a useful picture of the emission process, and also by methods which do not rely upon the Markov or atom-field statistical factorization assumption [2.2,14-16].

2.2.2 Collisional Relaxation

Under the conditions described in Sect.2.1.1, collisions may be described in the impact approximation, where they give rise to additional damping constants in the optical Bloch equations. If no energy-increasing ($|0\rangle \rightarrow |1\rangle$) collisions occur, the damping constants κ and κ' in the closed two-level case may then be represented as [2.17,18]

$$\begin{aligned} \kappa &= \Gamma + Q_I \\ \kappa' &= \frac{1}{2}(\Gamma + Q_I + Q_E) \quad , \end{aligned} \tag{2.13}$$

where Γ is the Einstein A coefficient (radiative width) for the transition, and Q_I and Q_E are the mean rates of inelastic (quenching) and elastic (dephasing) collisions, respectively.

The emission spectrum in this case may again be solved for in closed form [2.19], by using the method outlined in Sect.2.1.4. In the weak field limit, the solution is [2.17,18,20]

$$\tilde{g}(\nu) = \frac{\Omega^2}{4(\Delta^2 + \kappa'^2)} \left\{ 2\pi\delta(\nu - \omega) + \frac{Q_E}{\Gamma + Q_I} \left[\frac{2\kappa'}{(\nu - \omega_{10})^2 + \kappa'^2} \right] \right\} \tag{2.14}$$

and thus contains a broadened component proportional to Q_E which is centered at the atomic resonance frequency ω_{10}, in addition to the coherent component at ω. The spectral symmetry which is present quite generally in the radiative case is thus absent even in the weak field limit when elastic collisions occur.

In the case of well-separated spectral lines ($\Omega' \gg \kappa'$), the spectrum still has the form (2.10), and with widths still given by (2.12) [though with κ and κ' given by (2.13)], but the integrated-intensity coefficients are importantly modified by the collisional process. The sideband coefficients A_+ and A_- are in this case unequal, and are given by the relations [2.7]

$$A_\pm = \Omega^2(\Omega' \pm \Delta)\,[\eta(\Omega' \pm \Delta) \mp \Delta]/8\Omega'^2(\eta\Omega^2 + \Delta^2) \quad , \tag{2.15}$$

where

$$\eta \equiv \kappa'/\kappa = (\Gamma + Q_I + Q_E)/2(\Gamma + Q_I) \quad . \tag{2.16}$$

When the widths of the spectral lines are ignored, the spectrum appears as three sharp lines centered at ω, $\omega - \Omega'$, and $\omega + \Omega'$, with total intensities A_0, A_+, and A_-, respectively. The intensity of the central term (which actually consists of a coherent plus an incoherent part) is [2.7,21]

$$A_0 = \Omega^2/4\Omega'^2 \quad , \tag{2.17}$$

and hence, unlike the intensities $A_\pm$ of the sidebands, is completely independent of η, and thus of the *type* of relaxation mechanism.

The predictions of the collisional theory just outlined are in reasonable agreement with the experimental observations of CARLSTEN et al. [2.22].

2.2.3 General Solution for the Closed Two-Level System

When the relaxation mechanism (which can be due partly to the action of an incoherent broadband pumping field) includes energy-increasing processes (so that $\kappa_{01} \neq 0$) as well as energy-decreasing processes, the general solution for the emission spectrum for the closed two-level system is given in closed form by [Ref.2.19, Eqs. (2.5,11,16; 4.9)].

2.3 The Open Two-Level System

When relaxation-induced (spontaneous radiative or collisional) transitions take place from (or to) either of the laser-coupled states $|0\rangle$ or $|1\rangle$ to (or from) another, uncoupled state $|j\rangle$ of the atom (and thus when $|0\rangle$ need not be the ground state), generalizations of the methods described above are necessary to describe the emission spectrum both for the transitions in question and for the laser-coupled transition ($|1\rangle \rightarrow |0\rangle$) itself. These generalizations are entirely straightforward, as indeed are the ones which describe the case (not considered here) in which more than two states are directly coupled by the laser field. The detailed solutions in such cases generally depend upon all of the relaxation constants κ_{jk} and κ'_{jk} which connect any pair of states which are coupled directly or indirectly by the relaxation process to $|1\rangle$ or $|0\rangle$ (e.g., all of the states in a cascade from $|1\rangle$ to $|0\rangle$). In many cases, however, the solutions (or some of their limiting forms) can be expressed in terms of quantities which relate only to the transition in question, i.e., only the corresponding steady-state density matrix elements and/or certain other quantities with obvious physical significance (e.g., repopulation rates), which in turn can either be regarded as adjustable parameters or measured experimentally. (The assumption of steady-state conditions requires that repopulation into the $|1\rangle$ - $|0\rangle$ subspace take place if there is decay out of it).

2.3.1 Spectrum for Transitions Involving Other Levels

The emission spectrum for a radiative transition of a laser-coupled state $|1\rangle$ or $|0\rangle$ to or from another, uncoupled state $|j\rangle$ is given quite generally in terms of the steady-state populations of the pair of states involved and the single off-diagonal matrix element $\bar{\rho}_{10}$ by [Ref.2.23, Eqs. (3.13; 4.8,9)]. Typically two rather than three peaks appear in the spectrum in these cases. When the emission occurs during the transition from the state $|1\rangle$ to a state $|j\rangle$ of lower energy, for example, then in the limit of weak laser field intensity there will be, in addition to the normal sponataneous emission term centered at ω_{1j} (which however appears only if elastic collisions take place), a second term centered at $\omega_{1j} + \Delta = \omega - \omega_{j0}$ which is due to the Raman effect. As the laser intensity is increased, the separation of these two spectral terms increases to the Rabi frequency $\Omega' = (\Delta^2 + \Omega^2)^{\frac{1}{2}}$, while their mean value remains constant. In the strong field limit [Ref.2.23, Eq. (4.13)], the two terms are equal in integrated intensity and width, the latter being equal to $(\kappa'_{j1} + \kappa'_{j0})/2$.

2.3.2 Effect of Atomic Decay on the Laser-Coupled Spectrum

Relaxation-induced transitions out of the laser-coupled subspace $|1\rangle$ - $|0\rangle$ have effects, as mentioned above, on the emission spectrum for the $|1\rangle \rightarrow |0\rangle$ transition itself which in general depend upon the details of the transition sequence, and thus upon parameters relating to other states of the atom.

An important exception to this rule for which closed solutions are available occurs when the $|1\rangle$ - $|0\rangle$ subspace is repopulated not at all or else very slowly. In the former case, where no repopulation takes place, the $|1\rangle \rightarrow |0\rangle$ emission is not a steady-state phenomenon, but rather a transient one. The solution for the $|1\rangle \rightarrow |0\rangle$ spectrum in this case is given quite generally by [Ref.2.24, Eqs. (6-11)]. In place of the δ-function term which appears in the steady-state solution for the closed two-level case, a term appears which is broadened by the decay process. This term still represents a coherent effect provided that all of the atoms are actually prepared in the state $|0\rangle$ at the same initial time and the transient emission process is then observed, for all of the atomic dipole moments will oscillate in phase with one another even while the (Rabi-modulated) amplitude of the oscillations decays to zero.

If repopulation into the $|1\rangle$ - $|0\rangle$ subspace takes place, then a steady state will finally be reached no matter how small the repopulation rate R is. If R is small enough so that the steady-state probability of finding an atom in either of the states $|0\rangle$ or $|1\rangle$ is small compared to unity, then the steady-state emission spectrum for the $|1\rangle \rightarrow |0\rangle$ transition is directly obtainable [2.24] to lowest order in R from the exact solution for the transient emission process described in the preceding paragraph. In fact, that solution now describes the *incoherent* part of the spectrum, to first order in R. The "broadened δ function", in particular, represents an incoherent term under steady-state conditions in which the dipole oscillations for the $|1\rangle \rightarrow |0\rangle$ transition, though occurring with a prescribed (incident field-determined) phase, occur at different times for different atoms simply because of the limiting assumption $R \rightarrow 0$. The truly coherent term, which is always a strict δ function in steady state, does not appear directly in the solution under discussion because it vanishes to first order in R. It is, in fact, proportional to R^2, and its solution to lowest order is thus given by (2.6), where $\bar{\rho}_{10}$, which is linear in R, can be found directly from [Ref.2.24, Eqs. (11,14)]. Thus, for a sufficiently small repopulation rate, the solution for the steady-state emission spectrum for the $|1\rangle \to |0\rangle$ transition can be expressed entirely in terms of quantities relating to the transition in question.

When the repopulation rate is large enough, on the other hand, so that the probability of finding an atom in the $|1\rangle$ - $|0\rangle$ subspace is no longer infinitesimal, closed solutions for the $|1\rangle \to |0\rangle$ emission spectrum can only be expressed independently of the details of the decay-repopulation sequence in special cases. Notable among these is the limit of well-separated spectral lines, where the integrated intensities of all three components and the widths of the sidebands are given by [Ref.2.7, Eqs. (15,29)]. Also of interest are the closed solutions found by COOPER and BALLAGH [2.25] in the weak field limit, and the one-parameter solution in [Ref.2.7, Eq. (20)] for the case of a cascade process.

Supported by the National Science Foundation.

References

2.1 R.J. Glauber: Phys. Rev. *131*, 2766 (1963)
2.2 B.R. Mollow: Phys. Rev. A*12*, 1919 (1975)
2.3 B.R. Mollow: Phys. Rev. *188*, 1969 (1969)
2.4 M. Lax: Phys. Rev. *172*, 350 (1968)
2.5 B.R. Mollow, M.M. Miller: Ann. Phys. (New York) *52*, 464 (1969)
2.6 R.I. Sokolovskii: Zh. Eksp. Teor. Fiz. *59*, 799 (1970) [English transl.: Sov. Phys. - JETP *32*, 438 (1971)]
2.7 B.R. Mollow: Phys. Rev. A*15*, 1023 (1977)
2.8 F.Y. Wu, R.E. Grove, S. Ezekiel: Phys. Rev. Lett. *35*, 1426 (1975)
2.9 R.E. Grove, F.Y. Wu, S. Ezekiel: Phys. Rev. A*15*, 227 (1977)
2.10 H. Walther: "Atomic Fluorescence Induced by Monochromatic Excitation", in *Laser Spectroscopy*, Int. Proc. of the Second Conf., Megeve, France, 1975, ed. by S. Haroche, J.C. Pebay-Peyroula, T.W. Hänsch, S.E. Harris, Lecture Notes in Physics, Vol.43 (Springer, Berlin, Heidelberg, New York 1975) pp.358
2.11 W. Hartig, W. Rasmüssen, R. Schieder, H. Walther: Z. Phys. A*278*, 205 (1976)
2.12 H.J. Carmichael, D.F. Walls: J. Phys. B*9*, 1199 (1976)
2.13 C. Cohen-Tannoudji, S. Reynaud: J. Phys. B*10*, 345 (1977)
2.14 B.R. Mollow: J. Phys. A*11*, L130 (1975)
2.15 S. Swain: J. Phys. B*8*, L437 (1975)
2.16 H.J. Kimble, L. Mandel: Phys. Rev. A*13*, 2123 (1976)
2.17 D.L. Huber: Phys. Rev. *178*, 93 (1969)
2.18 A. Omont, E.W. Smith, J. Cooper: Astrophys. J. *175*, 185 (1972)
2.19 B.R. Mollow: Phys. Rev. A*5*, 2217 (1972)
2.20 B.R. Mollow: "Response Functions for Strongly Driven Systems", in *Coherence and Quantum Optics*, Proc. 3rd Rochester Conf., Rochester, New York, U.S.A., June 21-23, 1972, ed. by L. Mandel, E. Wolf (Plenum, New York 1973) pp.525-532
2.21 J.L. Carlsten, A. Szöke: Phys. Rev. Lett. *36*, 667 (1976)
2.22 J.L. Carlsten, A. Szöke, M.G. Raymer: Phys. Rev. A*15*, 1029 (1977)
2.23 B.R. Mollow: Phys. Rev. A*8*, 1949 (1973)
2.24 B.R. Mollow: Phys. Rev. A*13*, 758 (1976)
2.25 J. Cooper, R.J. Ballagh: Phys. Rev. A*18*, 1302 (1978)

3. Resonance Fluorescence of Atoms in Strong Monochromatic Laser Fields

J. D. Cresser[1], J. Häger[1], G. Leuchs[2], M. Rateike[2], and H. Walther[1,2]

With 28 Figures

The investigation of atomic resonance fluorescence has always been of special interest as a means for the determination of atomic parameters. In addition, information on the interaction mechanism between atoms and radiation can be obtained. In the standard fluorescence experiment the frequency distribution of the incident photons is larger than the natural width of the respective transition; as a consequence the correlation time in the photon-atom interaction is determined by the lifetime of the atoms in the excited state. With the development of lasers and especially of tunable dye lasers in recent years it became possible to study the case where the incident radiation has a spectral distribution which is narrower than the natural width. This corresponds to a correlation time of the incoming light wave which is much longer than the excited-state lifetime. In this chapter a survey of experiments on the resonance fluorescence of atoms in monochromatic laser fields will be given.

3.1 Overview

The interaction of laser light with atomic systems has received considerable theoretical and experimental attention over the past decade. Until the advent of the laser, light sources for spectroscopy consisted of ordinary spectral lamps excited by DC or RF discharges, and produced light having a very broad spectral width and, hence, very short correlation time, and a relatively low intensity. For such fields both the experimental and theoretical results are in general well understood. However, the development of the laser made available light sources which are sufficiently intense that an atomic (or molecular) transition can be very easily saturated. In addition, the lasers are highly monochromatic having a coherence time much greater than typical natural lifetimes of excited atomic states, and finally, tunable, making it possible to selectively excite particular atomic transitions. As might be expected, it has been found that many new and interesting phenomena are associated with such fields interacting with atomic systems.

The theoretical analysis of this new physical situation requires the use of techniques more general than those found adequate in the case of thermal fields. In the latter case the weakness of the atom-field interaction meant that perturbative

1 Projektgruppe für Laserforschung der Max-Planck-Gesellschaft, Garching

2 Sektion Physik Universität München, Garching

techniques were generally sufficient. These techniques were based on the assumption that the initial state of the atomic system was essentially unchanged by the interaction. However, as saturation can be easily achieved with an intense laser field, more general nonperturbative methods are required. Furthermore, for a highly coherent field, one cannot consider successive photon emission and absorption processes as being independent as it is now possible for an atomic system to undergo many such processes during the correlation time of the laser field, and hence phase memory effects cannot be neglected.

Although a wide range of problems both theoretical and experimental involving laser fields have been studied, attention here is confined to just one aspect: the interaction of intense monochromatic light with atomic systems, in which it is the properties of the fluorescent light (i.e., the light scattered by the atom) which is of principal interest.

The simplest such system is also one which has attracted an enormous amount of interest: the problem of theoretically and experimentally determining the spectrum of the fluorescent light radiated by a two-level atom driven by an intense monochromatic field. This is the situation that gives rise to the AC Stark effect in which, for sufficiently strong fields, it is found that the spectrum of the scattered light splits into three peaks consisting of a central peak, centered at the driving field frequency with a width $\Gamma/2$ (Γ^{-1} = Einstein A coefficient) and having a height three times that of two symmetrically placed sidebands, each of width $3\Gamma/4$ and displaced from the central peak by the Rabi frequency. In addition there appears a delta-function (coherent) contribution also positioned at the driving frequency. In the limit of strong driving fields, the energy carried by this last contribution is negligible compared to the three-peak contribution. This result was first predicted by MOLLOW [3.1] and subsequently by many others, using a variety of techniques [3.2-12], and which has been very well confirmed experimentally [3.13-17].

Research has not been confined to this simple model however. As the theoretical understanding of the effect was placed on firmer ground, investigation was extended to treat more complex situations. These include considering the effects of the incident field having a nonzero spectral width, examining the scattering from multilevel systems, and also treating the case in which many identical atoms take part in the scattering process. In the latter case, cooperative effects between the atoms considerably modify the properties of the fluorescent light, with bistable behavior being observed in the fluorescent light spectrum [3.18].

It is not only the spectral property of the fluorescent light that has come under investigation. The examination of the intensity correlation of the scattered field in the basic two-level atom has also attracted much attention since fluorescent light exhibits the property of photon antibunching [3.7,8,10]. Further, the total fluorescence intensity contains much of interest in the case of scattering by a multilevel atom. Of particular interest here is the study of the level crossing effect (the Hanle effect) when a static magnetic field shifts the two excited Zeeman sublevels in a three-level system.

In Sect.3.2 a review is given of the various theoretical treatments of the above problems. First (Sect.3.2.1) a review is given of work that has been done on the AC-Stark-effect problem in its basic form, where the calculation of the fluorescent light spectrum is of primary interest. In Table 3.1 a listing is given of the most important papers treating this problem. In Sect.3.2.2, still for this basic model, the various studies of the other properties of the scattered field, including the intensity, the intensity correlation, and the antibunching phenomena are reviewed. A summary of relevant papers is given in Table 3.2. Finally, in Sect. 3.2.3, a discussion is given of the various variants of the basic model, with once again the relevant publications listed in Table 3.3.

3.2 Theoretical Treatments of Interaction of Atoms with Intense Monochromatic Fields

3.2.1 Simple AC Stark Effect: Spectrum

The theoretical treatment of resonance fluorescence from a two-level atom irradiated by a monochromatic light field in the low-intensity limit, was first reviewed by HEITLER [3.19]. A scattered field spectrum was predicted which was very sharply peaked around the incident field frequency. The high-intensity limit was first considered by APANASEVICH [3.20] who, by numerical calculations based on earlier theoretical work [3.21], predicted a three-peak spectrum. Subsequently NEWSTEIN [3.22] also examined the problem with, however, collisional rather than radiation damping providing the relaxation mechanism. He also predicted a three-peak spectrum in the high-intensity limit, though due to a different damping mechanism the widths and heights of the three peaks differed from those later found in the pure radiation damping case. However, the first complete theoretical treatment in which exact expressions were obtained for the scattered field spectrum when radiation damping is present is the work of MOLLOW [3.1]. In his work the scattering atom was driven near resonance by a monochromatic classical electric field. The atom came into equilibrium with this field through the effects of radiation damping, this being included in the theory by explicitly coupling the atom to the quantized electromagnetic field. The solution was based on deriving the optical Bloch equations for the elements of the (2×2) reduced density matrix of the atomic system. These equations were obtained from a master equation approach in the derivation of which the Markov approximation was made. The diagonal elements of this reduced density matrix are just the probabilities of the atom being found in its ground or excited states, while the off-diagonal elements essentially give the mean dipole moment of the radiating atom.

However, it is not the mean dipole moment of the atom that acts as the source of the radiated field; rather it is the instantaneous value of the dipole moment, i.e., its mean value plus quantum fluctuations. This is recognized in MOLLOW's work in which, rather than calculating the correlation function of the mean dipole moment, and hence, by a Fourier transform, the spectrum of the radiated field, it is the correlation function of the dipole moment operator that is found so that quantum fluctuations are not averaged out. This latter correlation function is obtained from the optical Bloch equations by use of the quantum regression theorem [3.23].

Since MOLLOW used a classical description of the incident field, his method originally was not believed to be a fully quantum-electrodynamic treatment [3.24], although it was later shown by MOLLOW [3.3] that this work was in fact equivalent to such a description. Following the work of MOLLOW, STROUD [3.24] made the first attempt at deriving a solution for the case in which the incident field was described quantum electrodynamically.

Table 3.1: AC Stark effect

Author	Method	Remarks
MOLLOW [3.1]	Markovian master eq. (MME)	Classical driving field
OLIVER et al. [3.2]	MME	Fully QED
CARMICHAEL, WALLS [3.10]	MME	On-resonance only
MOLLOW [3.3]	Photon state analysis	Rigorous analysis of approx.'s
HASSAN et al. [3.4]	Heisenberg eqs. of motion	On-resonance only
SWAIN [3.5]	Continued fraction	Showed importance of photon interference effects
COHEN-TANNOUDJI [3.6,7]	Langevin eq.	
KIMBLE, MANDEL [3.8]	Heisenberg eqs. of motion	
WODKIEWICZ, EBERLY [3.9]	Heisenberg eqs. of motion	Non-Markovian, no use of quantum regression theorem
BALLAGH [3.11]	Photon state analysis	Dressed atom; showed importance of photon interference effects
CRESSER [3.12]	Photon state analysis	Photon interference discussed
SMITHERS, FREEDHOFF [3.25]	Photon state analysis	No coherent contribution to spectrum predicted
RENAUD et al. [3.26,27]	Heisenberg eqs. of motion	
STROUD [3.24]	Photon state analysis	One-photon approximation

STROUD's work was prompted as much by the need to avoid the semiclassical approach of MOLLOW as by a desire to compare the results of a QED calculation with those obtained from the so-called neoclassical theory [3.28] in which the concept of a quantized electromagnetic field was avoided altogether. It was found later that the predictions of the QED approach were fully vindicated by experiment.

STROUD's work introduced the "dressed atom" method later popularised by COHEN-TANNOUDJI [3.6,7]. This method amounts to making a judicious choice of basis states, these states being eigenstates of the coupled atom-driving field system. The energy eigenvalue spectrum of the dressed atom system assumes the form of a series of doublets, the frequency separation between the members of a doublet being just the (off-resonance) Rabi frequency $\Omega' = (\Omega^2 + \Delta^2)^{\frac{1}{2}}$ where Ω is the on-resonance Rabi frequency and Δ is the detuning of the driving field away from resonance. The frequency separation between corresponding levels in successive doublets is just the frequency of the driving field. The scattering of photons can then be visualised as a sequence (or cascade) of spontaneous decays down through the states of the dressed atom system. STROUD, however, truncated this problem by considering only a single spontaneous transition and obtained a result similar to MOLLOW's, but differing from MOLLOW's results as regards the widths and relative heights of the three peaks. In the real physical situation there is a cascade through the successive energy levels of the dressed atom system accompanied by the spontaneous emission of many photons, and in a correct calculation of the spectrum, proper account must be taken of these photon cascades. Unfortunately this direct approach is very difficult since, among other problems, quantum interference effects associated with the different possible order of emission of the radiated photons must be properly included in the calculations. Such explicit photon descriptions of the AC Stark effect will be discussed later. The first fully QED treatment was not based on explicitly following the photon cascades. A Markovian master equation approach was used in which the reduced density operator for the dressed atom system was obtained [3.2]. This reduced density matrix is of a far more complex form than the 2×2 matrix obtained by MOLLOW [3.1] as the dressed atom system consists of a large number of states. OLIVER et al. used the quantum regression theorem to obtain the spectrum of the fluorescent field. However, no explicit expressions for this spectrum were reported, only computed plots of the spectrum were given. CARMICHAEL and WALLS [3.10] also made use of the dressed atom picture for a driving field exactly on resonance to obtain, by a fully QED method, expressions for the spectrum in agreement with MOLLOW's results.

Alternative fully QED treatments have also been given subsequent to CARMICHAEL and WALLS's work [3.4,8,9,26,27]. These methods were based on the use of a Heisenberg equation of motion approach in which the equations of motion of the atomic and field operators are obtained from the Hamiltonian of the total system and, by eliminating unwanted variables, are reduced to equations involving only atomic and free field operators. The atomic operators in these equations evolve (in the Heisenberg picture) under the action of the total Hamiltonian of an atom plus fields plus interaction, while the free field operators evolve under the Hamiltonian of the free field only. HASSAN and BULLOUGH [3.4] actually derive equations of motion for the atomic operators averaged over the initial state of the field, taken to be given by a coherent state, while in the other treatments this approximation was avoided. Nevertheless, results in complete agreement with MOLLOW's were obtained. In the above Heisenberg equation of motion approaches, in almost all cases a Markovian-type approximation was made in the derivation of the equations (an exception is [3.9]), although it was referred to by different names, the adiabatic approximation or the harmonic approximation. A claim by KIMBLE and MANDEL [3.8] that such an approximation was not made in their work was shown by ACKERHALT [3.29] to be in error.

COHEN-TANNOUDJI [3.7] made use of a Langevin equation of motion approach. In this method, the optical Bloch equations which are linear differential equations relating the mean values of atomic system operators for the two-level atom, were formally replaced by operator equations, with delta-function-correlated random force operators added to each equation to take account of quantum fluctuations. This is in accordance with an approach to quantum noise problems developed by LAX [3.30]. COHEN-TANNOUDJI obtained the usual (i.e., MOLLOW) result for the spectrum.

All these above methods relied on making a Markov approximation or atom-field statistical factorization assumption. The validitiy of these approximations were open to question [3.24] so that an approach to the problem was required which either avoided the approximations or else rigorously established their validity. To do this it was necessary to turn to the more difficult problem of working directly with the photon cascades. The first such method was developed by MOLLOW [3.3]. In this paper he assumed a coherent state description for the incident field and was able to show by a canonical transformation that this fully QED description of the model was exactly equivalent to one in which the field was treated classically, with the initial state of quantized field transformed into the vacuum state. Thus the equivalence of the QED approach and his original semiclassical approach [3.1] was rigorously established. Moreover, within the context of a well-defined set of approximations he then showed how all the photon reabsorption processes could be allowed for, leading to a new (non-Hermitean) Hamiltonian in which the energy of the upper state was assigned an imaginary part—its natural linewidth. The interaction term in this Hamiltonian only creates photons so that under the action of this Hamiltonian, the transformed initial state, i.e., the vacuum state, evolves into a linear combination of Fock states containing multiphoton contributions of all orders. Thus the photon cascade effect mentioned above was fully included in the theory. Of course, the usual expression for the spectrum was obtained.

An important aspect of this work is that the validity of all approximations made was very carefully established. MOLLOW was able to establish, therefore, the validity of the Markov approximation made in other approaches to this problem, and was also able to derive the quantum regression theorem, thereby placing its use in calculating the spectrum on firm ground. WODKIEWICZ [3.3] has investigated the consequences of not making the Markov approximation, and was able to show that a very slight asymmetry was to be expected in the scattered light spectrum, though the effect was shown to be very small and difficult to observe.

Other calculations of the spectrum based on working directly with the photon states were also performed by SMITHERS and FREEDHOFF [3.25], SWAIN [3.5], BALLAGH [3.11], and CRESSER [3.12], though MOLLOW in [3.3] did question the correctness of the method used in [3.25]. SWAIN based his calculations on his continued fraction method [3.32]. BALLAGH used the dressed atom picture and a Feynman diagram technique, while CRESSER made use of a generalization of the formalism of MOWER [3.33]. Of interest is the fact that these methods all showed the importance of quantum interference effects in determining the final form of the spectrum, and in fact it was shown that the coherent (delta-function) contribution to this spectrum is entirely due to interference effects. In this regard, BALLAGH gave a very complete description of how this coherent contribution builds up through the succession of spontaneous decays in the dressed atom picture.

3.2.2 Simple AC Stark Effect: Total Scattered Intensity, Intensity Correlations, and Photon Antibunching

Further information about the AC Stark effect can be obtained by examining properties of the scattered light field other than the spectrum. In Table 3.2 a list is given of various publications in which other aspects of the simple AC Stark effect problem are investigated.

The calculation of the total intensity of the scattered field was done in greatest detail by KIMBLE and MANDEL [3.8]. They gave computer plots of the intensity of the scattered field as a function of time for arbitrary driving field intensities and detunings. They showed that the scattered field intensity exhibited oscillatory behavior that became more apparent at high driving field intensities and increased detuning. The intensity was also shown to always settle down to a constant steady-state value after a sufficiently long time had elapsed.

The intensity correlations are of far more interest, however, than the intensity itself. The intensity correlation function was investigated in [3.8,10]. CARMICHAEL and WALLS [3.10] considered the case of an on-resonance driving field only, while [3.7,8] provided a generalization for arbitrary detuning of this field away from resonance. The importance of this work is that the intensity correlation was found to exhibit a behavior which has no classical counterpart.

Table 3.2. Antibunching

Author	Remarks
CARMICHAEL, WALLS [3.10]	Predicted antibunching phenomenon
KIMBLE, MANDEL [3.8]	Predicted antibunching phenomenon
COHEN-TANNOUDJI [3.7]	Predicted antibunching phenomenon
JAKEMAN et al. [3.34]	Included effects of atomic beam fluctuations
KIMBLE et al. [3.35]	Included effects of atomic beam fluctuations
CARMICHAEL et al. [3.36]	Included effects of atomic beam fluctuations
DAGENAIS, MANDEL [3.37]	Included effects of atomic beam fluctuations

For usual light fields (e.g., thermal light) it is found that the intensities of the field at two neighbouring instants in time are strongly correlated, i.e., if a photomultiplier irradiated by this field emits an electron at some instant in time, the probability is high for a second photoemission to occur a short time later. This phenomenon is known as photon bunching, and can be explained using either a classical or quantum-mechanical description of the light field.

However, it was found that for the field scattered by a single two-level atom, the intensity correlation function was of a form that showed that if a photon was detected (by a photomultiplier) at some instant in time, then the probability of detecting another photon during a short time interval following the first detection remained close to zero. This phenomenon is the reverse of that described earlier and is known as photon antibunching. This behavior can be explained quantum mechanically by the fact that the process of detecting a scattered photon also prepares the scattering atom in its ground state. Thus no further photons can be emitted and hence detected until the atom has had sufficient time to be pumped back up to its excited state (the only state from which emission can occur) by the driving field. It should be pointed out that fields exhibiting antibunching can be generated in other ways, i.e., by multiphoton absorption in which two or more photons are simultaneously absorbed [3.38,39], or else as a result of nonlinear optical effects in the degenerate parametric process first discussed by STOLER [3.40] and developed further by PAUL and BRUNNER [3.41], BANDILLA and RITZE [3.42, 43]. However, we will confine our attention here to antibunching in the case of resonance fluorescence only.

The significance of antibunching is that there does not exist a classical field which exhibits this behavior. Thus the existence of photon antibunching can be taken as a test of the validity of QED. However, the theoretical model on which the above result is based is not directly representative of a true experimental situation. All experimental studies of the AC Stark effect and related phenomena involve a beam of atoms passing perpendicularly through a laser field. However, the antibunching effect can be washed out if a number of atoms are simultaneously interacting with the laser field [3.10]. Thus, ideally, to observe the phenomena, the intensity of the atomic beam must be sufficiently low that only a single atom at a time passes through the field. In the real situation, however, even for low beam intensities, there is a statistical fluctuation in the number of atoms in the field at any time. Thus, for the purposes of comparing theory and experiment, the above theory must be modified to allow for fluctuations in the number of atoms in the field at any time, and also, as it turns out, the effect of the finite transit time of the atoms through the field must be also accounted for [3.34-36]. In addition the possible effects of nonzero laser bandwidth need to be examined, the pertinent work in this case being that of WODKIEWICZ [3.44] who used a phase diffusion model (PDM) of the laser field to calculate improved expressions for the intensity correlation. These corrections were found to be important at low laser field intensities (Sect. 3.3.3). Detailed examination of laser bandwidth effects have also been made for both the PDM and chaotic field model for the incident light [3.45,46].

Although differences in detail were found in these investigations, the antibunching phenomenon was still found to be present and detectable. It is found that the results of experiment [3.47,48] are in agreement with the improved theory. The experiments will be discussed in some detail in one of the following sections.

3.2.3 Variants of the AC Stark Effect

Theoretical investigation has also been conducted into variants of the simple form of the AC Stark effect problem. Table 3.3 lists a representative sample of the papers in which such variations of the basic problem have been studied.

Table 3.3. Variants of the basic model

Author	Variant considered
AGARWAL [3.49]	Nonzero spectral width of driving field
EBERLY [3.50]	Nonzero spectral width of driving field
AVAN, COHEN-TANNOUDJI [3.51]	Nonzero spectral width of driving field
KIMBLE, MANDEL [3.52]	Nonzero spectral width of driving field
ZOLLER [3.53]	Nonzero spectral width of driving field
KNIGHT et al. [3.54]	Nonzero spectral width of driving field
SOBOLEWSKA [3.55]	Three-level atom
SOBOLEWSKA, SOBOLEWSKI [3.56]	Three-level atom
KORNBLITH, EBERLY [3.57]	Three-level atom
AVAN, COHEN-TANNOUDJI [3.58]	Hanle effect
COHEN-TANNOUDJI, REYNAUD [3.59]	Multilevel atom
CARMICHAEL, WALLS [3.18]	N identical scattering atoms

RENAUD et al. [3.27], using a fully quantum-electrodynamic treatment developed in an earlier paper [3.26], investigated the spectrum of the scattered field detected during an observation period of finite length. One of the important results of their investigations was that for a weak driving field tuned off-resonance the spectrum of the transient field, i.e., the field radiated at the start of the atom-field interaction was asymmetric with the sideband closest to the atomic transition frequency enhanced. This transient behavior was later found to play an important role in determining the spectrum for a nonzero bandwidth driving field.

A number of publications have been devoted to examining the effects of a driving field having a nonzero bandwidth [3.49-54, 60-66]. In all these treatments, the nonzero bandwidth of the driving field was introduced by supposing that either the phase (PDM) or both the phase and amplitude of the field were subject to random fluctuations. AVAN and COHEN-TANNOUDJI [3.51] and AGARWAL [3.49] both examined the on-resonance case and showed that the spectrum still had a symmetric three-peak structure with, however, each of the peaks broadened. KIMBLE and MANDEL [3.52] used a generalized version of the Heisenberg operator technique developed in an earlier publication [3.8], which enabled them to treat both the on- and off-resonance situations. They showed that for an off-resonance driving field the scattered spectrum became markedly asymmetric, an effect noted in some experimental work. GEORGES and DIXIT [3.66] considered a more exact model for the phase fluctuations, taking proper account of their non-Markovian character as predicted by the theory of HAKEN [3.67]. KNIGHT et al. [3.54] were able to explain physically the origin of this antisymmetric structure. They used a simple Lorentz model appropriate to a weak driving field and used a method from EBERLY [3.50] to take account of the random nature of this field. They were able to show that the scattering atom cannot settle down to a steady-state phase relation with the fluctuating applied field. The atom continually falls out of phase with the field and is repeatedly returned to its transient interaction regime. As shown by RENAUD et al. [3.27] it was just in this transient regime that an asymmetrical spectrum was to be expected. However, in contrast to the transient effect discussed there, which was a once only affair associated with the turning on of the atom field interaction, the transient effect described by KNIGHT et al. [3.54] was an intrinsic property of the driving field and existed independently of the turn on of the field. They also showed that in the transient scattered field, it was the nonelastic contribution at the transition frequency of the atom due to the overlap of the excitation spectrum with the atomic absorption line that produced the enhanced sideband.

Other variations of the basic problem involve considerably more complex models for the scattering atom, e.g., COHEN-TANNOUDJI and REYNAUD [3.59] investigated the case of a multilevel atom acting as the scattering center, and used this theory to examine resonance Raman scattering in very intense fields [3.68]. The particular case of a three-level atom has been examined in [3.55-57]. The expressions for the spectra are very complex. For instance, KORNBLITH and EBERLY [3.57] obtained a seven-peak spectrum.

The study of the total fluorescent intensity radiated by a three-level atom is perhaps of more interest, in certain cases, than of the spectrum of the light. An example of this kind is the study of the total fluorescent light when a static field is scanned around the value corresponding to a crossing between two excited sublevels (level crossing or Hanle effect). The theory of the Hanle effect for monochromatic excitation has been developed by AVAN and COHEN-TANNOUDJI [3.58].

The Hanle effect is well known in classical experiments involving broadband excitation of the atomic states, but a fundamental difference occurs when the atom is irradiated by monochromatic light. This is best seen by examining the elements of the reduced density matrix of the atomic system which now contains nonzero off-diagonal elements coupling the excited and ground states (optical coherences), thus representing a coherent superposition of the ground and excited states. For broadband excitation, these coherences are completely washed out. However, as the optical coherences are not negligible for monochromatic excitation, more complex behavior is expected in this case than in the case of broadband excitation. A detailed discussion is given below of the experimental investigations of the Hanle effect for monochromatic excitation. A detailed comparison with the theoretical results of [3.58] is also given.

3.3 Experiments on the Interaction of Atoms with Intense Monochromatic Fields

In the following section the experiments on the resonance fluorescence induced by intense monochromatic radiation will be reviewed. In Sects.3.3.2,3 some so far unpublished results will be described.

3.3.1 Emission Spectrum

The spectrum of the scattered light is, as discussed above, related to the Fourier transform of the first-order correlation function of the atomic operators. We shall summarize the theoretical results as follows: For low laser intensities the atom remains very close to its ground state and behaves like a classical oscillator (see also Sect.3.3.2). The light is therefore scattered elastically, and for a monochromatic driving field one observes a sharp spectrum at the same frequency as the driving field (Fig.3.2). As the intensity of the exciting light increases the atom spends more time in the upper state and the effect of the vacuum fluctuations due to spontaneous emission comes into play. An inelastic component enters the spectrum, and the magnitude of the elastic scattering component is correspondingly reduced. The spectrum gradually broadens as the Rabi frequency Ω increases until Ω exceeds

$\Gamma/4$; then sidebands begin to appear. For the saturated atom the form of the spectrum shows three well-separated Lorentzian peaks. The central peak has the width $\Gamma/2$ and the sidebands which are displaced from the central peak by the Rabi frequency are broadened to $3\Gamma/4$. The ratio of the height of the central peak to the sidebands is 3:1.

Experimental study of the problem requires that Doppler broadening be almost completely excluded. Therefore the laser light has to be scattered by the free atoms of a strongly collimated atomic beam. In order to measure the frequency distribution, the fluorescent light has to be investigated by means of a highly resolving spectrometer. The first experiments of this type have been performed by STROUD et al. [3.13] and later by WALTHER et al. [3.14,16] and by EZEKIEL et al. [3.15,17]. In all experiments the excitation was performed by single-mode dye lasers and the scattered radiation was analyzed by Fabry-Perot interferometers.

In the following the experiments of [3.16] will be discussed in more detail. First results obtained for the fluorescence spectrum at low laser intensities will be described; in the second part the high-intensity results are reviewed.

The scheme of the experimental set-up is shown in Fig.3.1. The atomic beam was collimated by circular apertures with a collimation ratio of about 1:500. This corresponds to a residual Doppler width of about 2 MHz. The direction of the atomic beam, the axis of excitation, and observation were mutually perpendicular. The interaction region between atomic beam and laser light was inside a confocal Fabry-Perot.

Using this arrangement the fluorescence signal is enhanced by a factor which is almost equal to the finesse of the interferometer. The apertures A_1 and A_2 defined an acceptance angle for the multiplier of about 2.5 mrad. This opening angle corresponds to a residual Doppler width of about 3 MHz when the atomic beam is adjusted to be perpendicular to the axis of the Fabry-Perot. The total linewidth observable in the experiment is determined by the collimation ratio of the atomic beam, the opening angle for observation, the laser linewidth, and the finesse of the Fabry-Perot. With the numbers given above a total linewidth of about 10 MHz had to be expected.

For the study of the frequency distribution of the fluorescent light at low light intensities the transition $6s^2\ {}^1S_0 - 6s6p\ {}^1P_1$ of ^{138}Ba was used. This transition has the wavelength $\lambda = 5535$ Å. It has been chosen since its natural linewidth is 20 MHz which is twice as large as the width of the sodium D_2 line. As the expected total linewidth for the measurement of the spectrum was 10 MHz, this resolution should be sufficient to demonstrate the frequency distribution of the fluorescence to be smaller than the natural width. Experiments at larger laser powers have not been performed on Ba as the power required to separate the side maxima from the main signal component would have been twice as large as for the corresponding measurements on the Na D_2 line.

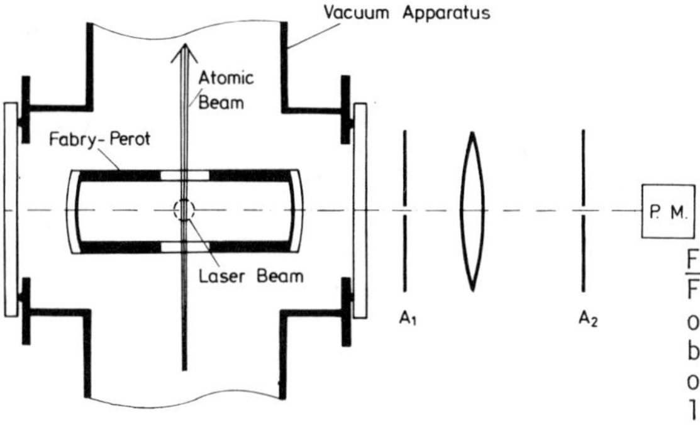

Fig.3.1. Experimental setup. The Fabry-Perot had a free spectral range of 300 MHz. The interferometer could be tuned by a piezoelectric variation of the cavity length. The finesse was larger than 100

Figure 3.2 shows a result for the Ba transition. The dashed curve represents a Lorentzian with the natural width of the transition; the solid curve shows the observed spectrum for the fluorescence. The measured halfwidth is about 12 MHz. Similar results have also been obtained by other authors on the corresponding transition of Mg [3.69]. The natural width of this line is 80 MHz; therefore a smaller resolution is required than for the Ba transition to demonstrate the fluorescence spectrum to be sharper than the natural width.

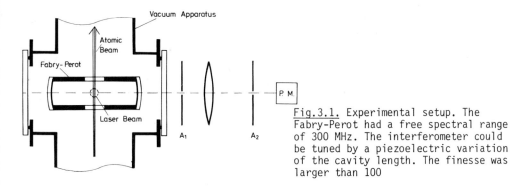

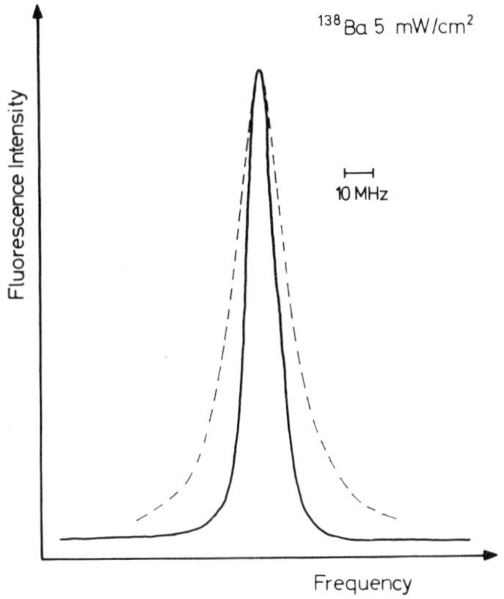

Fig.3.2. Spectrum of the fluorescent light of the ^{138}Ba resonance line at 5535 Å. The excitation was performed at the center of the transition with a laser power of 5 mW. The dashed line is the Lorentzian with the natural width of 20 MHz

The experiments with high laser intensity were performed at the F'=3-F=2 hyperfine transition of the sodium D_2 line. The hyperfine structure of the D_2 line is shown in Fig.3.3. This transition is suitable as the upper F'=3 level can only decay into the F=2 level of the ground state from where the excitation is performed; therefore multiple excitations are possible and, in addition, no hyperfine pumping can occur. The transition has, of course, the disadvantage that it deviates from the two-level system usually considered in the theoretical treat-

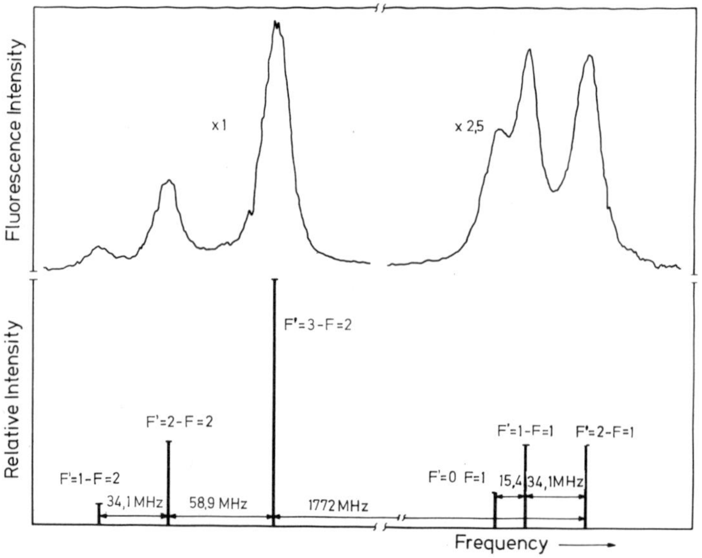

Fig.3.3. Hyperfine splitting of the Na D_2 line. For this measurement the total fluorescent light of an atomic beam (without frequency selection) was observed by suitably sweeping the CW dye laser frequency. The frequency scale is interrupted between the two groups of components belonging to different F. The lower part shows the theoretical structure. F' and F denote the quantum numbers of the total angular momentum of the hyperfine levels of the $^2P_{3/2}$ and $^2S_{1/2}$ levels, respectively

ments as the two hyperfine levels are degenerate. Furthermore the other hyperfine levels of $^2P_{3/2}$ are so close that their influence cannot be neglected at high laser intensities, as they overlap with the F'=3 level due to power broadening. As a consequence spontaneous decay to the $^2S_{1/2}$, F=1 can occur besides the decay to the F=2 level. Since the hyperfine splitting between F=1 and F=2 is 1772 MHz, a reexcitation of those atoms is impossible.

The degeneracy of the hyperfine levels has an influence on the frequency distribution of the fluorescent light since the Rabi nutation frequency which determines the distance of the side components depends on the transition probability; therefore the different Zeeman substates contribute in a different way to the spectrum. In order to reduce this effect the number of Zeeman states involved has to be reduced. This is possible by using circularly polarized light for excitation. Due to optical pumping during the first excitation processes when the atoms enter the laser beam, almost all are pumped into the $m_{F'}=2$ or $m_{F'}=-2$ Zeeman sublevel of the F=2 hyperfine level for right-handed or left-handed circular polarization, respectively. Finally only the transition $^2S_{1/2}$, F=2, $m_{F'}=\pm 2 \rightarrow {}^2P_{3/2}$, F'=3, $m_F=\pm 3$ is excited and no other Zeeman transition has to be taken into account. In addition, no coupling with other $^2P_{3/2}$ hyperfine levels is possible since there is no other allowed transition starting from $^2S_{1/2}$, F'=2, $m_{F'}=\pm 2$ which may be populated by the laser light; as a consequence the two-level system is closely approached [3.70].

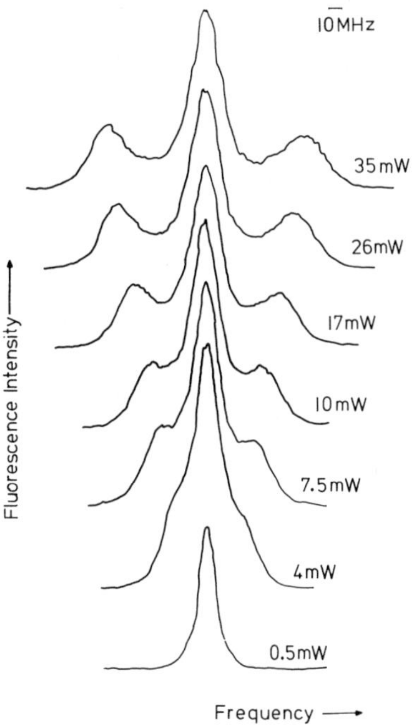

Fig.3.4. Spectra of the fluorescent light from the F'=3-F=2 hyperfine transition for different laser powers

For the experiments two single-mode CW dye lasers of the folded cavity type were used [3.71,72]. The linewidth of both lasers was less than 1 MHz when measured with a Fabry-Perot interferometer in a time of about 30 s. One laser was stabilized to the $^2P_{3/2}$, F'=3 → $^2S_{1/2}$, F=2 transition of sodium using a separate atomic beam. The second laser was free running with an output power of up to 60 mW in a single mode. The second laser beam was heterodyned with the first one in order to determine the frequency of the second laser with respect to the atomic transition frequency. The beat signal was analyzed with a radio frequency spectrum analyzer. Thus the difference frequency between the two lasers could be accurately determined.

Figure 3.4 shows the spectra of the fluorescent light for different laser powers. The laser frequency was tuned to the center of the resonance line. It is evident that the positions of the side maxima vary with the laser power. The signal intensity of the side components is about one third of that of the central peak. This is in agreement with the predictions of the theory of MOLLOW [3.1] and others (Sect. 3.2.1).

The separation of the two side maxima from the central component is, for an excitation on resonance according to MOLLOW, given by the Rabi nutation frequency

$$\Omega = \sqrt{\omega^2 - (\Gamma/4)^2} \quad ,$$

where Γ is the natural decay constant of the atomic transition and $\omega^2 = 4|\mu \underline{e}|^2 |E_0|^2/\hbar^2$, where μ is the value of the electric dipole moment of the transition, $\underline{e}$ the polarization vector of the electric field, and E_0 the amplitude of the electric field. The separation of the side maxima deduced from the experiment is in very good agreement with theory [3.16].

The dependence of the position of the side maxima on the detuning of the laser with respect to the accurate transition frequency is shown in Fig.3.5. The laser power was 30 mW. For a larger detuning the side maxima move further away from the central component and become smaller. In [3.16] the distance between the side and central components was studied in detail as a function of the detuning. This comparison gives an excellent agreement, too.

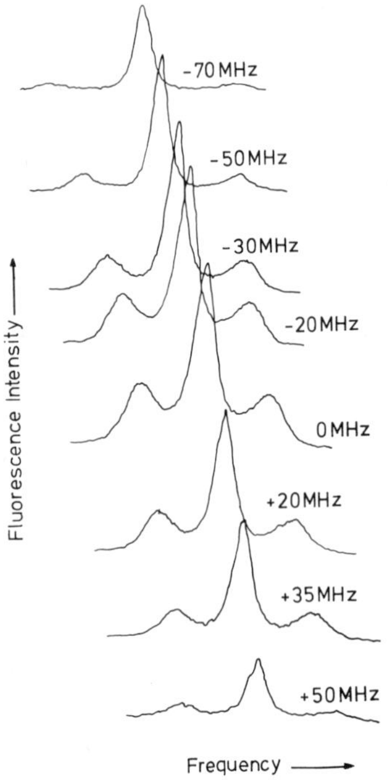

Fig.3.5. Variation of the spectrum of the fluorescence for different detunings of the laser frequency with respect to the transition frequency. The laser power for all measurements was 30 mW

In [31.6,17] the convolution of the theoretical spectra with the instrumental line shape (linewidth about 10 MHz in both experiments) was also compared to the measured spectra. There is a good agreement. However, one small discrepancy between the measurements of EZEKIEL et al. [3.17] and our results should be discussed briefly. The experimental results show under some conditions asymmetries, i.e., one side peak is smaller than the other. In [3.16] results are presented which indicate that the use of linearly polarized light leads to asymmetry due to the possibility of exciting transitions to more than one Zeeman substate. However, EZEKIEL et al. [3.17] demonstrated that asymmetric spectra may be produced by observing atoms in a nonuniform field, regardless of polarization. Hence, misalignment may be responsible for the observed deviation. It is quite probable that there are con-

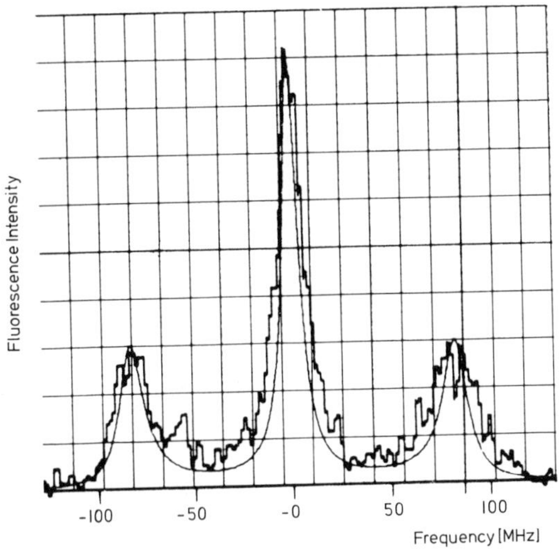

Fig.3.6. Theoretical and experimental spectra from strongly driven F'=3-F=2 hyperfine transition of Na. The Doppler broadening of the beam has been reduced using a lower oven temperature. The theoretical spectrum is shown as a smooth curve [3.17]

tributions by both effects; however, these small discrepancies are of no principal importance. The agreement between theory and experiment can be considered to be established in the essential features.

For completeness one very nice result of [3.17] should still be presented here (Fig.3.6). This spectrum was taken at a very low temperature of the atomic beam oven. The spectrum shows a reduction in the width of all three peaks (about 17 MHz for the central peak and 32 MHz for the side peaks) due to a smaller Doppler broadening. The measured spectrum is compared directly with the theory represented in Fig. 3.6 by the smooth curve for Ω = 82 MHz. The fit is reasonably good if it is considered that the remaining Doppler broadening and the field inhomogeneities of the laser field in the excitation region have not been included.

3.3.2 Total Emitted Intensity: Level Crossing Experiments

The interaction of monochromatic laser radiation with atoms is characterized by a correlation time of the light wave much longer than the respective excited-state lifetime of the particles. Therefore, each atom undergoes many absorption and emission processes during the correlation time of the laser wave. In the two-level situation this interaction can be described by Bloch-type equations. Many phenomena in laser physics have been explained in this way, the representation being especially useful in connection with the description of coherent transient effects.

The optical Bloch equation includes the spontaneous decay of the excited states as relaxation of the average dipole moment. This gives, in general, an adequate description. However, when the frequency distribution of the spontaneously emitted light is investigated, not only the mean variation of the average dipole is important, but also its fluctuations around the mean. This has been discussed in the previous section.

When the spontaneously emitted radiation is observed without frequency selection the above-mentioned fluctuations of the electric dipole are of no importance and the experiments are again well described by Bloch-type equations. An example of this kind is the study of the level crossing resonance which appears on the total fluorescent light when a static field is scanned around the value corresponding to a crossing between two excited sublevels (level crossing or Hanle effect [3.73]). In the case of monochromatic excitation the density matrix description of this effect involves not only the nondiagonal elements which describe the coherent superposition of the sublevels but also the nondiagonal elements connecting the fundamental state and the excited state. Under monochromatic excitation a coherent oscillation between the excited and fundamental states is built up, being analogous to the Rabi nutation in magnetic resonance. The nondiagonal elements of the optical transition are of course of no importance in the case of broadband excitation since the short coherence time of the radiation does not allow a coherent mixing of excited and fundamental states. Therefore, the level crossing experiment under monochromatic excitation gives some information not contained in classical experiments.

a) *Level Crossing Experiment*

First results on level crossing experiments with monochromatic excitation were published some time ago [3.74]. The $6s^2\ {}^1S_0$-$6s6p\ {}^1P_1$ resonance transition in the BaI spectrum at 5535 Å was investigated. These measurements showed the influence of monochromatic excitation on the level crossing signal form only qualitatively. In a more recent experiment [3.14,75] it was possible for monochromatic excitation to determine the dependence of the form of the level crossing signal on both the laser intensity and laser frequency detuning. This allowed a comparison to be made with theoretical signal forms calculated for a ${}^1S_0$-${}^1P_1$ transition [3.58]. In order to achieve excitation of the atoms by the same laser frequency the Doppler effect had to be eliminated. Therefore the light was scattered by the free atoms of a well-collimated atomic beam (collimation ratio about 1:500). This corresponds to a residual Doppler width of about 2 MHz. The absorption width of the Ba atomic beam was therefore essentially determined by the natural width of the ${}^1P_1$ level being 20 MHz. For the experiments Ba with natural isotope mixture was used. The laser was tuned to the ^{138}Ba line to avoid the complication of hyperfine splitting. This hfs of the 5535 Å Ba lines [3.76,77] with their relative fluorescence intensities is shown in Fig.3.7. The dye laser used in the experiment has been described earlier [3.72]. The spectral linewidth of the free running laser is less than 1 MHz, with a frequency drift of less than 1 MHz/min. The maximum output power at 5535 Å was about 10 mW.

The experimental geometry was chosen so that the magnetic field was directed parallel to the atomic beam. The direction of the exciting light, the direction of the magnetic field, and the direction of observation were mutually perpendicular. The laser light was linearly polarized perpendicular to the magnetic field. The part

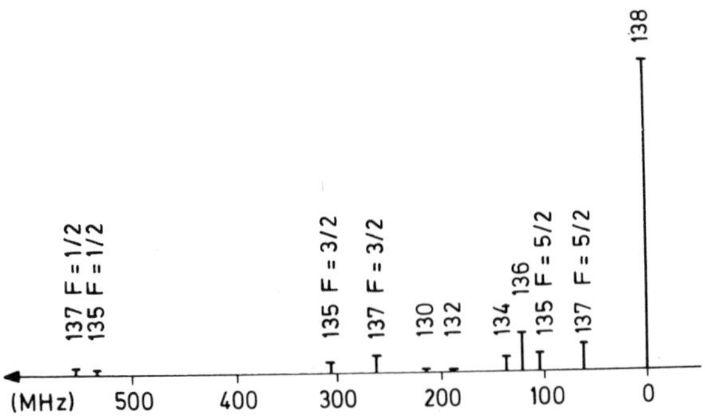

Fig.3.7. Hyperfine splitting and isotopic shift of the Ba 5535 Å line

of the fluorescent light polarized perpendicular to the magnetic field was observed.

In the standard level crossing experiment this experimental geometry gives a variation of the fluorescence light described by a Lorentzian with a minimum at zero magnetic field and maximal values at high positive and negative magnetic field values. In the case of monochromatic excitation this signal is superimposed by the decreasing fluorescence intensity due to the magnetic scanning of the Zeeman levels.

The atomic beam diameter was about 0.2 mm, the laser beam diameter about 2 mm. By the use of an aperture it was possible to observe just the fluorescence light resulting from a central part of the interaction volume (nearly homogeneous density of laser power). The fluorescence light L_x was detected by means of a photomultiplier, amplified, and then fed into a voltage-to-frequency converter. The pulses were then stored in a multichannel analyzer (MCA) in the multiscaling mode (Fig. 3.8). The magnetic field was swept according to the channel variation of the MCA. The Earth's magnetic field was compensated by using auxiliary field coils.

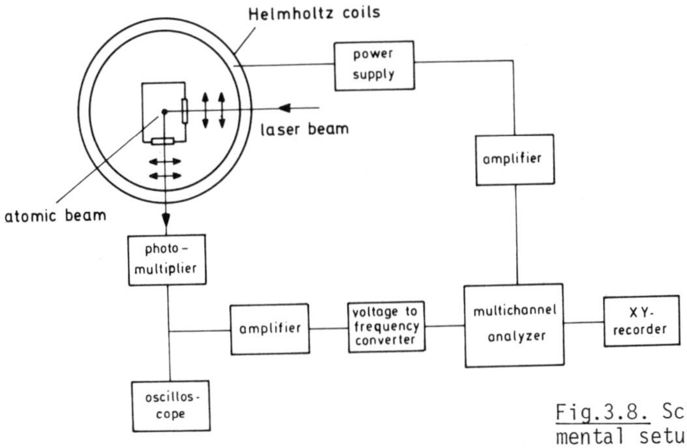

Fig.3.8. Schematics of the experimental setup

b) *Theoretical Considerations*

As the other chapters of this book do not deal with the theoretical description of the level crossing effect under monochromatic excitation, this will be done briefly in this section.

The ^{138}Ba $^1S_0-^1P_1$ transition near 5535 Å is a nearly ideal transition for measuring level crossing signal forms. The level scheme for the experiment is shown in Fig.3.9. Because of the excitation by σ light only three magnetic sublevels have to be considered: $M_J = \pm 1$ for the excited state and $M_J = 0$ for the ground state. After the coherent excitation the atom can be described by the superposition

$$|\psi\rangle = \alpha|\psi_{+1}\rangle + \beta|\psi_{-1}\rangle$$

where $|\psi_{\pm 1}\rangle$ are the excited magnetic substates, and α and β describe their time dependence.

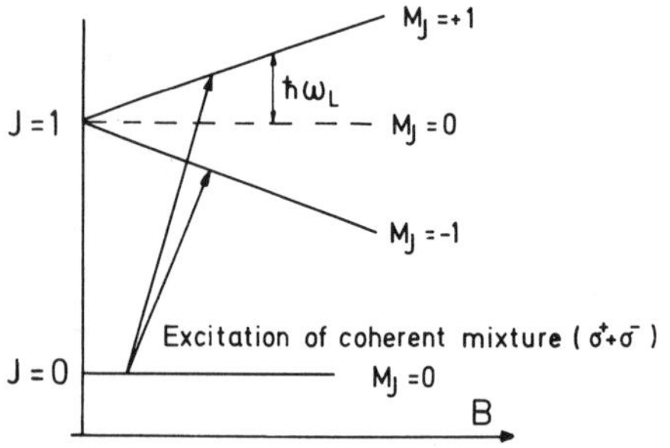

Fig.3.9. Principle of the level crossing experiment. J = 0 → J = 1 transition and excitation by σ light

For vanishing magnetic fields the coherence produced during excitation is conserved and the reemitted radiation shows an angular distribution due to the linear polarized excitation (vanishing intensity on the detector for our experimental geometry). With an increasing magnetic field, however, the energy separation of the states $|\psi_{+1}\rangle$ and $|\psi_{-1}\rangle$ also increases. As a consequence, the phase difference between the oscillatory parts of the coefficients α and β also increases and the coherence decreases. For the particular geometry of the experiments this means that the observed fluorescence intensity gets larger. This effect is similar in monochromatic as in broadband excitation. In the case of monochromatic excitation, however, the signal decreases at high magnetic fields, a result of the magnetic scanning effect caused by the Zeeman splitting of the levels, as is shown in Fig.3.10. The signal around the zero field is mainly influenced by the interference of the transition amplitudes, whereas the scanning of the magnetic sublevels determines the signal form for higher magnetic fields. For the calculation of the correct signal form, the equations which describe the evolution of the atomic density matrix have to be solved [3.58].

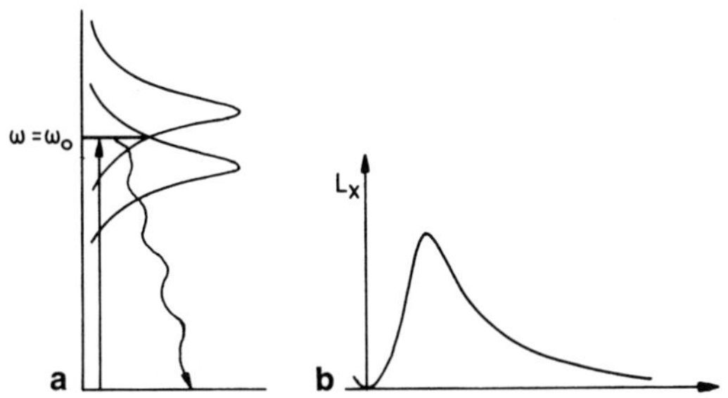

Fig.3.10a,b. Level crossing signal with monochromatic excitation

$$\frac{d}{dt}\rho = \frac{1}{i\hbar}[\underset{\sim}{H},\rho] + F(\rho) \quad . \tag{3.1}$$

The density matrix for the problem looks as follows:

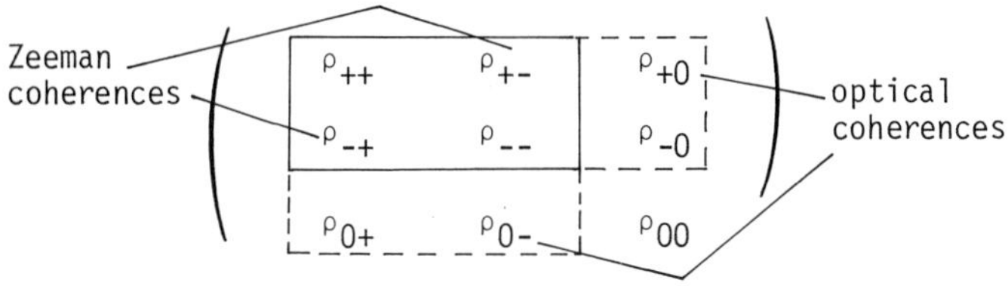

$\underset{\sim}{H}$ is the total Hamiltonian of the system including the interaction with the exciting beam, and $F(\rho)$ describes the effect of spontaneous emission. As the coherence time of the light is long compared to the lifetime of the excited state, as well as to the nondiagonal elements ρ_{+-},ρ_{-+} of the density matrix describing the Zeeman coherences in the excited state, the optical coherences $\rho_{\pm 0},\rho_{0\pm}$ between the excited and ground state also have to be considered. These nondiagonal elements represent the motion of the electric dipole moments driven by the incident laser light.

For small pumping intensities (far below the saturation intensity) the signal form is given by the square of a Lorentzian,

$$L_x \propto \frac{\omega_L^2}{(\omega_L^2 + \Gamma^2/4)^2} \tag{3.2}$$

where L_x is the fluorescence intensity in the above-described experimental geometry, ω_L is the Larmor frequency, and Γ the natural linewidth. One factor of the denominator is determined by the decrease of the Zeeman coherence due to the Larmor precession and the other by the detuning between laser frequency and the two Zeeman levels with increasing magnetic field. The evaluation of the level crossing signal form depending on pumping power and detuning for the present geometry was performed in [3.58] and yielded

$$L_x \propto \frac{\omega_L^2(D-1) - \omega_L Y}{A + (3D-1)\cdot B} \tag{3.3}$$

where

$$x^{\pm} = [(\Gamma/2)^2 + (\Delta \pm \omega_L)^2]v^{-2} \quad ,$$

$$D = 1 + (1 + x^+)^{-1} + (1 + x^-)^{-1} \quad ,$$

$$Y = (\Delta - \omega_L)(1 + x^-)^{-1} - (\Delta + \omega_L)(1 + x^+)^{-1} \quad ,$$

$$A = 4\omega_L(Y + \omega_L D) + 3Y^2$$

$$B = \frac{\Gamma^2}{4}\left(2 + \frac{1}{x^+} + \frac{1}{x^-}\right) + \frac{(\Delta - \omega_L)^2}{x^-(1 + x^-)} + \frac{(\Delta + \omega_L)^2}{x^+(1 + x^+)} \quad ,$$

$$\Delta = \omega - \omega_0 \quad ,$$

ω is the pumping laser frequency, ω_0 the resonance frequency, $v^2 = 3E^2e^2f_{0\pm}/16m\omega_0$ is proportional to the square of the Rabi nutation frequency, and $f_{0\pm}$ is the oscillator strength of the two transitions. The parameter $4v^2/\Gamma^2$ is proportional to the laser intensity, i.e., it determines the power broadening in relation to the natural linewidth.

The result for the given experimental geometry, for the case of zero detuning ($\Delta = 0$) and for all intensities is given by

$$L_x \propto \frac{\omega_L^2 v^2}{[(\Gamma/2)^2 + \omega_L^2]^2 + 4v^4 + [5(\Gamma/2)^2 + \omega_L^2]v^2} \tag{3.4}$$

It is evident that (3.2) follows from (3.4) for vanishing power broadening ($4v^2/\Gamma^2$ very small).

It should be mentioned that recently AVAN and COHEN-TANNOUDJI have extended their study of the level crossing signal to the case of a fluctuating intense quasi-monochromatic laser beam [3.51]. Assuming that the laser intensity is so high that several Rabi nutations occur during the correlation time of the laser light and that the atoms are sensitive to the laser fluctuations during their lifetime (spectral width of the laser larger or comparable to the decay constant) the level crossing signal was calculated. These conditions do not allow a treatment by means of Bloch equations or by rate equations which are used in the case of broadband excitation. A light wave was assumed which has short-time fluctuations resulting from an erratic motion of the phase and long-time fluctuations which may by caused by a slow phase diffusion or a slow amplitude variation. The short-time fluctuations were considered by a perturbation treatment and the slow variations were assumed to be adiabatically followed by the atom. AVAN and COHEN-TANNOUDJI have shown in this way that the level crossing signal is quite sensitive to fast phase fluctuations and that a detailed study of the signal shape of the level crossing signal could give an interesting insight into higher order correlations of the light beam.

c) *Measurements*

The fluorescence intensity of the excited ^{138}Ba atoms was measured versus magnetic field for a fixed laser frequency. In a first set of experiments the laser was tuned exactly to the ^{138}Ba transition frequency (zero detuning); in a second set the laser was stabilized at fixed frequencies besides that of the ^{138}Ba transition ($\Delta \neq 0$).

All the measurements were performed with natural Barium, that means with a natural mixture of Ba isotopes. Therefore it is important to consider the possible signal influence of all the Ba isotopes with their isotopic shifts as well as their hyperfine and Zeeman splittings. As is shown in Fig.3.7 four hfs lines are only up to 150 MHz away from the ^{138}Ba transition frequency, so that they can influence the level crossing signals, especially at high excitation intensity (power broadening).

Figure 3.11 shows two parts of a high-resolution Ba spectrum of the 5535 Å line. The fluorescence intensity is plotted versus the frequency of the laser exciting the Ba atoms for very low pumping power. The ^{138}Ba line is well resolved and shows its natural linewidth. Relevant for level crossing measurements are the lines in Fig. 3.11b; all other components cannot influence the signals because of their large separation from the ^{138}Ba transition.

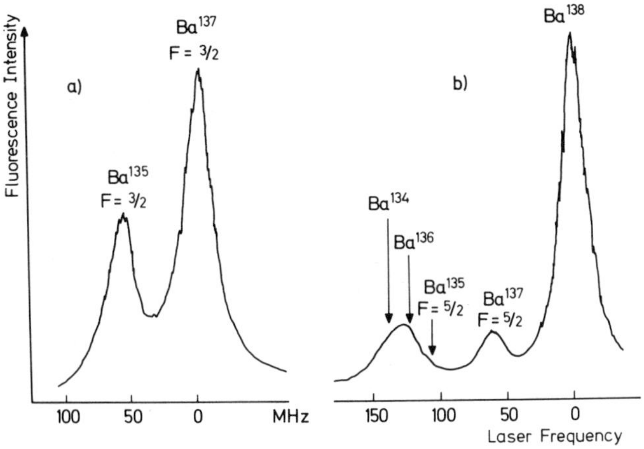

Fig.3.11a,b. High-resolution spectra of some hfs transitions of the 5535 Å Ba line at very weak excitation intensity. The fluorescence intensities of (a) and (b) are not reproduced in scale

Due to the power broadening effect the transitions in Fig.3.11b overlap more and more with increasing laser intensity. This effect is demonstrated in Fig.3.12 showing the Ba hfs near the ^{138}Ba line at a laser power of about 1.5 mW (the same frequency region as in Fig.3.11b). As can be seen the fluorescence intensity of the ^{138}Ba line is increased by contributions from neighboring transitions. Of course this effect has to be taken into account when the level crossing signal forms at higher laser intensities are interpreted.

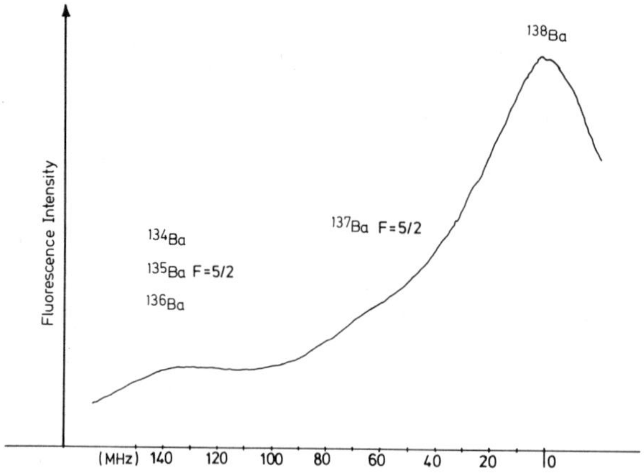

Fig.3.12. High resolution spectrum near the ^{138}Ba transition frequency at an excitation power of 1.5 mW. Overlapping of the various lines due to power broadening

For the case of zero detuning (laser frequency exactly on the ^{138}Ba line) magnetic scanning brings the two Zeeman components out of resonance, but simultaneously some Zeeman components of the other isotopes come into resonance. The theoretical Zeeman tuning of the hfs lines near the ^{138}Ba transition is displayed in Fig.3.13. The positions of the relevant Zeeman sublevels are plotted versus the magnetic field. As can be seen, components of unwanted Ba isotopes come into resonance with the laser frequency at magnetic fields of about 50 and 90 Gauss. At these magnetic fields the disturbing components show a fluorescence which overlaps the normal level crossing signal of the ^{138}Ba transition. The frequency region of this superposition is of course dependent on the power broadened linewidth.

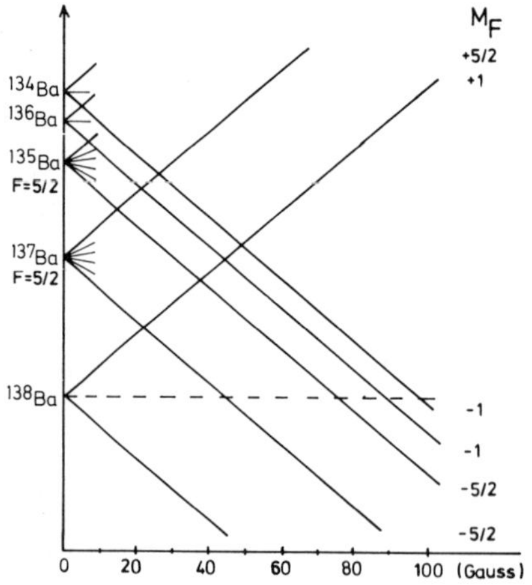

Fig.3.13. Zeeman tuning of the sublevels near the ^{138}Ba transition frequency

The Zeeman scanning is demonstrated in Fig.3.14 by experimental fluorescence signals at various magnetic fields. The fluorescence intensity is plotted versus the frequency around the original ^{138}Ba transition indicated by the broken line. The

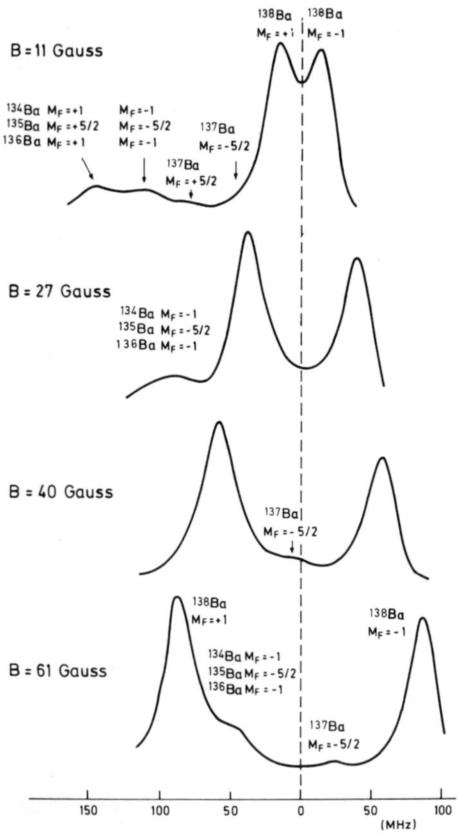

Fig.3.14. High resolution fluorescence spectra at different magnetic fields (very small laser power)

^{138}Ba components as well as all the other isotopic lines show magnetic scanning. It is obvious that the unwanted lines cross the original ^{138}Ba frequency at several magnetic fields, as was predicted according to Fig.3.13.

At small magnetic fields and weak laser intensities the level crossing signal form should only slightly be influenced by the Zeeman tuning of the hyperfine components; at higher intensities, however, power broadening and Zeeman splitting should strongly influence the reemitted fluorescence light.

The measured level crossing curves with monochromatic excitation are discussed in the following. Figures 3.15-18 show the results for the case of zero detuning between laser frequency and transition frequency. The fluorescence intensity is displayed versus magnetic field and the Larmor frequency. The curves are normalized to the same height. The parameters given on the figures are the values $4v^2/\Gamma^2$, proportional to the laser intensity. The solid points are the theoretical values given by (3.4).

Figure 3.15 shows a level crossing signal measured at an excitation power of about 5 μW, an intensity far below the saturation intensity. It is obvious that a theoretical curve (3.3,4) with an intensity parameter $4v^2/\Gamma^2$ = 0.015 agrees very

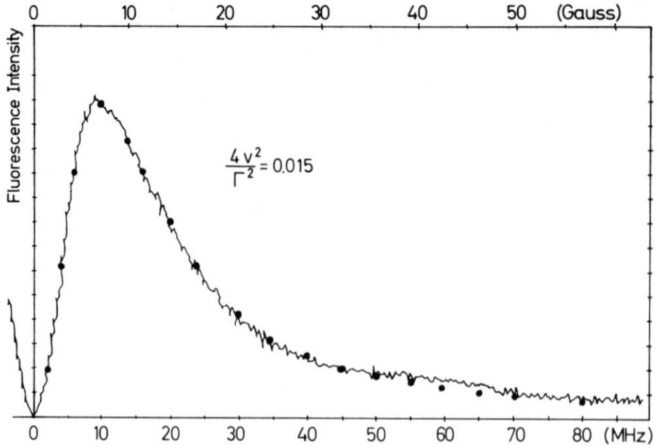

Fig.3.15. Level crossing signal at an excitation power of about 5 μW. The dots are theoretical values. The parameter is the fitted $4v^2/\Gamma^2$ value which is proportional to the laser intensity

well with the experimental signal. Also the halfwidth of the dip at zero magnetic field is about 45% of the halfwidth of a usual level crossing signal, as should be expected according to (3.2). The small deviation on the slope of the curve near 60 Gauss is the predicted contribution of the ^{137}Ba $M_F = -5/2$ component to the fluorescence intensity.

An increase in laser power broadens the level crossing curves, which can be seen by comparing Figs.3.15,16. The signal in Fig.3.16 was measured at an excitation intensity of about 60 μW. Theoretical and experimental curves agree very well for an intensity parameter $4v^2/\Gamma^2 = 0.2$. The contribution to the slope is the same as in Fig.3.15.

Increasing the excitation intensity to 1.3 mW yields further broadening, as can be seen in Fig.3.17 (intensity parameter $4v^2/\Gamma^2 = 3.4$, compared to a curve with $4v^2/\Gamma^2 = 0.2$). At intensity parameters $4v^2/\Gamma^2 > 1$, however, it is not possible to fit the experimental curves directly, without any correction. At such high excitation intensities the power broadened lines of the unwanted Ba isotopes even influence the fluorescence intensity at vanishing magnetic fields.

A careful evaluation of this contribution at different magnetic fields yields corrected level crossing curves, indicated by the dashed signal shape in Figs.3.17, 18 (2.2 mW). These corrected signals can successfully be fitted and show good agreement. The small discrepancies for the highest intensity can be attributed to uncertainties of the correction and to the possibility that contributions to the fluorescence can come from a part of the interaction volume which saw a smaller excitation intensity. Another reason for this residual discrepancy could be a fluctuating laser frequency within the laser bandwidth. As already mentioned, AVAN and COHEN-TANNOUDJI calculated such an influence of a fluctuating laser beam on the

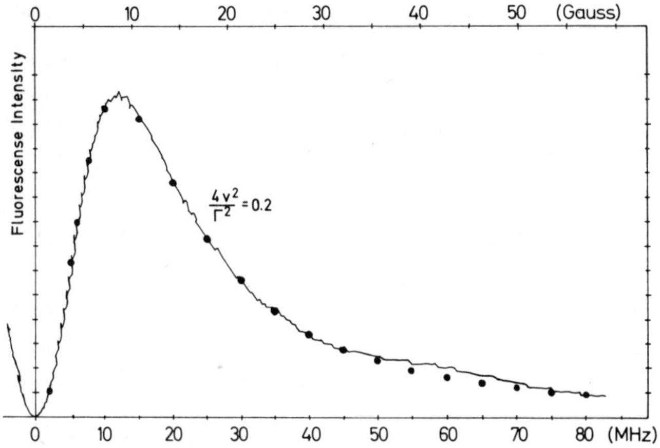

Fig.3.16

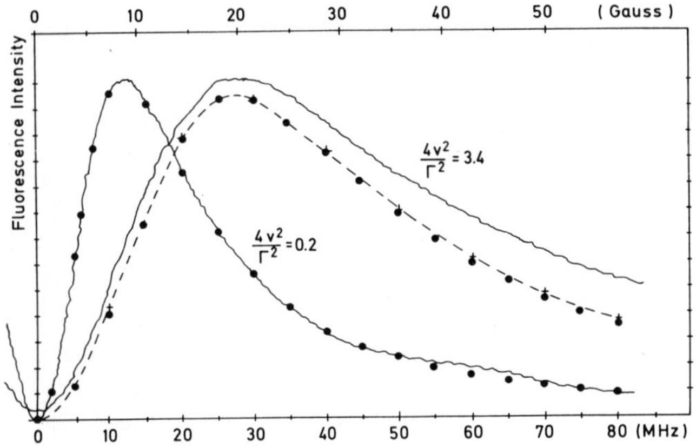

Fig.3.17

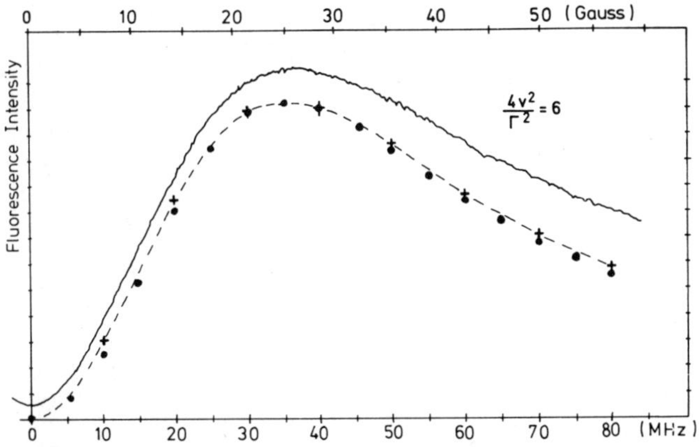

Fig.3.18

Fig.3.16, Level crossing signal at an excitation power of about 60 μW

Fig.3.17. Comparison between two level crossing signals with 60 μW and 1.3 mW excitation intensity. The broken line indicates the level crossing signal after subtraction of the contributions from the perturbing transitions

Fig.3.18. Level crossing signal at an excitation power of 2.2 mW

level crossing signal form, however only for a very high excitation intensity and a fluctuation linewidth which is broader or equal to the natural linewidth. The results of these calculations showed a deviation toward a smaller linewidth at small magnetic fields and a broadening at higher magnetic fields, compared to signal forms obtained with monochromatic excitation. Qualitatively the fit in Fig.3.18 shows the same deviations.

Equation (3.3) also contains signal forms which have to be expected when the laser frequency does not exactly agree with the atomic resonance frequency ($\Delta \neq 0$). In order to check these results, measurements with selected detunings of the laser frequency have also been performed. The experimental curves are displayed in Figs. 3.19,20. The parameters indicate the detunings in MHz, the numbers in parentheses represent the intensity parameters $4v^2/\Gamma^2$, and the dots indicate the theoretical values from (3.3). The measurements in Figs.3.19,20 were performed with a laser power of 0.35 mW and 0.6 mW, respectively. As it is rather difficult to keep the pumping intensity exactly constant during the frequency tuning, small deviations in the intensity parameters occur.

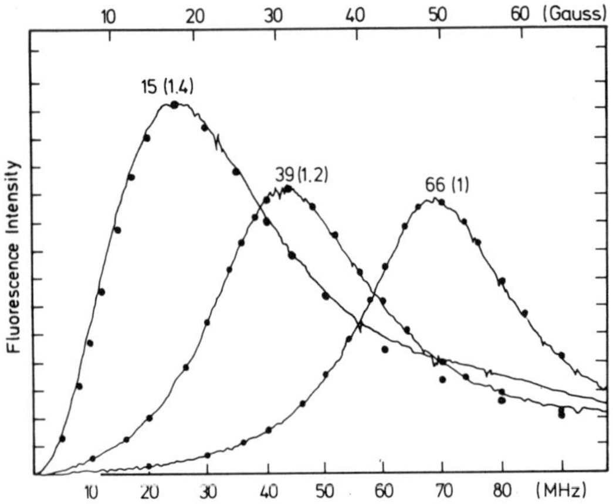

Fig.3.19. Level crossing signals for different detunings between laser frequency and ^{138}Ba transition frequency. The parameters give the detuning in MHz; the numbers in parentheses are the intensity parameters $4v^2/\Gamma^2$. (Excitation power: 0.35 mW)

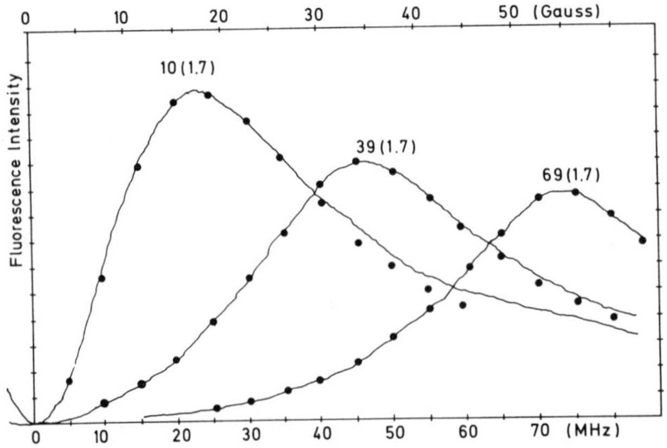

Fig.3.20. Level crossing signals for different detunings at an excitation power of 0.6 mW

The behavior of the curves is similar to that described for the signals with zero detuning; the influence of the unwanted Ba isotopes increases with increasing laser power, especially for small detunings [e.g., the values 15 (1.4) and 10 (1.7) in Figs.3.19,20, respectively]. In this case a correction seemed not to be necessary. The experimental curves also show good agreement with theory without correction.

3.3.3 Intensity Correlation: Photon Antibunching

Further information on the nature of the fluorescent light may be obtained via the second-order correlation function of the light defined by

$$g^{(2)}(\tau) = \frac{\langle E^{(-)}(t)E^{(-)}(t+\tau)E^{(+)}(t+\tau)E^{(+)}(t)\rangle}{[\langle E^{(-)}(t)E^{(+)}(t)\rangle]^2}$$

where $E^{(+)}(t)$ and $E^{(-)}(t)$ are the positive and negative frequency components of the electromagnetic field, respectively. The second-order correlation function was introduced by GLAUBER [3.78] in his formulation of optical coherence theory. In essence, $g^2(\tau)$ is a measure for the probability that a second photon will be measured at time $t + \tau$ in a light beam after the detection of one at time t. The result of a photon correlation experiment [3.79] for a chaotic light source with Gaussian frequency distribution (e.g., discharge lamp) is a Gaussian with a peak at $\tau = 0$ and a width given by the inverse bandwidth of the light source. That means that there is a tendency for the photons to arrive in pairs, or in bunches.

If instead of a discharge lamp, a highly stabilized laser is used for the correlation experiment, one obtains a constant $g^{(2)}(\tau)$. This result holds even when the laser and discharge lamp have the same bandwidth. Hence there is some fundamental difference between a laser and a chaotic light source which may not be apparent in the spectrum or the first-order correlation function which may, however, be seen in the second-order correlation function.

The second-order correlation function of the light in resonance fluorescence has been calculated by CARMICHAEL and WALLS and others [3.7,8,10]. The result in the steady state for the saturated atom ($\Omega \gg \Gamma$) is

$$g^{(2)}(\tau) = (1 - e^{-3\Gamma\tau/4} \cos \Omega\tau) \quad .$$

We see that this function exhibits damped oscillations at the Rabi frequency. Moreover, the extraordinary feature of this correlation function is that it begins at zero and increases. This is quite unknown in electromagnetic fields produced by classical sources. For example, the $g^{(2)}(\tau)$ for chaotic light fields as measured in the Hanbury-Brown and Twiss experiment begins at two and decreases to one. For coherent light such as produced by ideal lasers $g^{(2)}(\tau)$ has a constant value of one.

The function $g^{(2)}(0)$ can be interpreted as a measure of the probability for photons to arrive in pairs. Hence for chaotic light where $g^{(2)}(0)$ is twice the random background, this effect has been termed "photon bunching". The behavior evident in $g^{(2)}(\tau)$ for resonance fluorescence is a manifestation of quite the opposite effect, i.e., "photon antibunching." It is known that the quantum theory of the electromagnetic field allows for such behavior. If we evaluate the correlation function via a probability distribution $P(\mathscr{E})$ for the complex field amplitude $\mathscr{E}$ we find (see, e.g., [3.48])

$$g^2(0) - 1 = \frac{\int P(\mathscr{E})(|\mathscr{E}|^2 - <|\mathscr{E}|^2>)^2 d^2\mathscr{E}}{<|\mathscr{E}|^2>^2} .$$

Classically $P(\mathscr{E})$ is a probability distribution and $g^{(2)}(0)-1$ is always positive ensuring $g^{(2)}(0) \geqslant 1$. For a chaotic field with a Gaussian distribution for $P(\mathscr{E})$ we find $g^{(2)}(0) = 2$, whereas for a coherent field with a stabilized amplitude $P(\mathscr{E})$ is a delta function and hence $g^{(2)}(0) = 1$. Therefore the photon correlation for chaotic fields and for coherent laser fields is adequately described by classical theory. Since $g^{(2)}(0) \geqslant 1$ must be fulfilled in the classical case, antibunching is not allowed.

However, in the quantum formulation of the electromagnetic field $P(\mathscr{E})$ is a quasi-probability distribution and may take on negative values allowing for a $g^{(2)}(0) < 1$. The interpretation of this effect in the phenomenon of resonance fluorescence goes as follows: The first photon detected implies the atom has undergone an emission process and so is now in its ground state. In order to register a correlation, a second photon must be detected. It is clear that there must be a time lapse for the atom to regain its excited state. In fact, $g^{(2)}(\tau)$ is just proportional to the probability of observing the atom in the upper state when it was initially prepared in the ground state (3.3). We note that as the probability for the atom being in the upper state increases, $g^{(2)}(\tau)$ may exceed 1, and we see photon bunching and photon antibunching exhibited in the same phenomenon. In order to observe the photon antibunching one requires a very small number of atoms in the observation region (preferably $\lesssim 1$). In the presence of a large number of atoms the heterodyne signal from the beating of light from different atoms will completely obscure the photon antibunching in the light from a single atom.

In a dilute atomic beam with a mean number of atoms in the observation region $\lesssim 1$, one still has the problem of atomic number fluctuations. As pointed out by JAKEMAN et al. [3.34] the correlation measurement will include the statistics of the atomic number fluctuations. For Poissonian number fluctuations this implies that $g^{(2)}(0) = 1$. The photon antibunching then cannot be directly observed but is superposed on the atomic number fluctuations. It has been suggested [3.77] that an alternative normalization scheme of the correlation function which reduces the effect of the atomic number fluctuations may enable photon antibunching with a $g^{(2)}(0) < 1$ to be observed directly.

So far two photon correlation experiments have been performed on Na atoms in order to observe the photon antibunching. The first one has been published by MANDEL and coworkers [3.35,37,47], and the second one has been carried out in our laboratory. As the latter results have not yet been published in detail [3.48] this experiment will be described in the following.

The dye laser used in the experiment was the same as that employed for the measurements of the fluorescence spectrum (Sect.3.3.1). The cavity length had to be stabilized to compensate for frequency drifts, since signal averaging up to two hours was necessary in the experiment. The standard way of stabilizing a laser is to modulate the cavity length. This results in a modulation of the output frequency, which is a disadvantage for the experiment as the effective linewidth of the laser is broadened in this way. Therefore the resonance frequency of the reference atomic beam was modulated instead. The reference atomic beam with a collimation ratio of 1:500 was exposed to an oscillating magnetic field produced by a pair of Helmholtz coils (Fig.3.21). Part of the dye laser beam having an intensity low enough not to saturate the atomic transition was circularly polarized and directed at right angles onto the reference atomic beam, inducing the $F = 2 \rightarrow F' = 3$ hyperfine transition of the D_2 line. Because of the different g factors of the two levels the atomic transition frequency was modulated by the oscillating magnetic field. As a result the atomic fluorescence was also modulated. Using the standard lock-in technique a feedback signal was derived to stabilize the cavity length. With the magnetic field oscillating around $\langle B \rangle = 0$, the laser was stabilized exactly on resonance. By introducing an offset $\langle B \rangle \neq 0$ the laser could also be stabilized to frequencies being slightly off resonance.

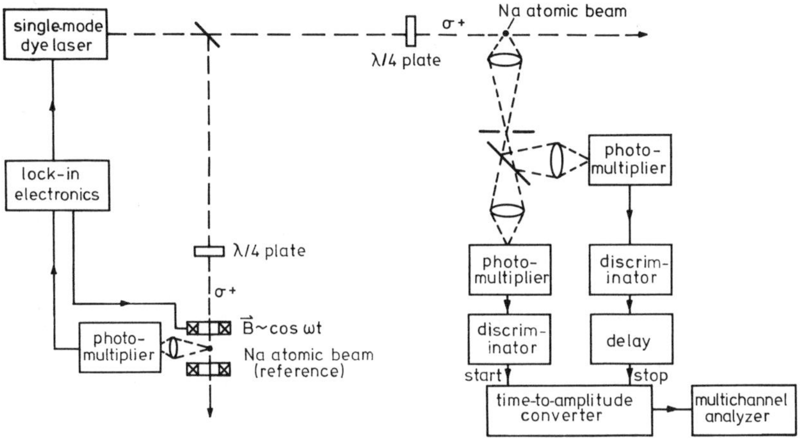

Fig.3.21. Experimental setup for the photon correlation experiment

For the study of the photon correlation of the fluorescence, the frequency-stabilized dye laser beam is directed onto the highly collimated atomic sodium beam (collimation ratio 1:1000) at right angles (Fig.3.21). Using circularly polarized light the Zeeman sublevels of the atoms are optically pumped and after about 100 spontaneous emissions only the $3^2S_{1/2}$, F = 2, m_F = 2 level is populated which then can be coupled to the $3^2P_{3/2}$, F' = 3, m_F = 3 level. Thus the sodium atoms represent a two-level system to a good approximation, as discussed in Sect. 3.3.2.

In order to be able to observe Rabi oscillations in the photon correlation it is essential that the atoms in the observation region are subject to a constant laser intensity. The spatial distribution of the laser beam, however, is described by a Gaussian profile. Therefore, the observation has to be limited to the maximum of the Gaussian profile, only allowing for laser intensity changes of <10%. In addition, the laser beam is expanded by taking advantage of the divergence of the dye laser, so that the Gaussian beam profile is enlarged. Using a collecting lens the fluorescing part of the atomic beam is magnified by 5:1 and imaged onto an aperture of 1 mm diameter, which transmits only the center part of the fluorescence coming from the maximum of the Gaussian profile. The atoms have an average thermal velocity of $8 \cdot 10^4$ cm/s. As a result the transit time of the atoms through the observation region is about 250 ns. The wings of the Gaussian laser profile, the fluorescence of which is not observed, serve to optically pump the atoms. By the time the atoms enter the observation region, they form two-level systems.

In the experiment the time interval between subsequent fluorescence photons had to be measured. A time-to-amplitude converter (TAC) was used to obtain the ns time resolution. The TAC converts two fast pulses (start and stop) into one slow (μs) pulse the amplitude of which is proportional to the time interval between the start and the stop pulses. In order to avoid false signals produced by photomultiplier ringing, two separate photomultipliers were used for detection. The fluorescent light is divided by a beam splitter. The photomultiplier pulses were shaped by a discriminator and then used to start and stop the TAC.

The output of the TAC was processed in a multichannel analyzer (MCA) working in the pulse height analysis mode. The channel number in which a TAC pulse is stored corresponds to the time interval τ between start and stop pulse. After averaging, the data in the MCA represent the probability that two subsequent photomultiplier pulses are separated by the time interval τ, as a function of τ. The count rates were so low that no pile-up effect has been observed. Thus, the averaged data in the MCA directly give the second-order correlation function.

The stop pulse was delayed by 38 ns, so that pairs of photomultiplier pulses corresponding to photons arriving at the same time on the cathodes of the two multipliers were counted as if they were separated by 38 ns. Thus, $\tau = 0$ is shifted to a higher channel number and the second-order correlation function is also measured

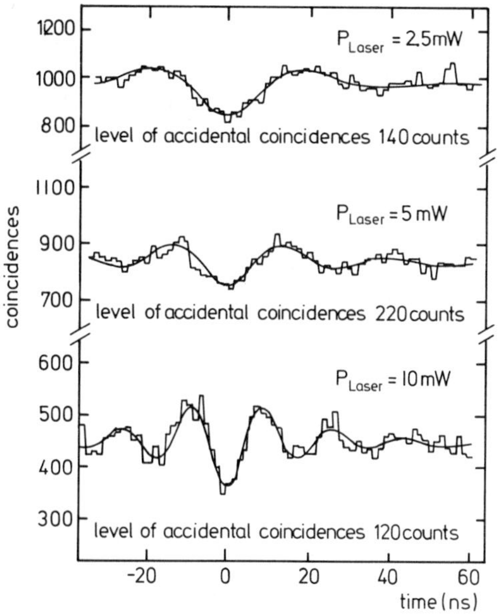

Fig.3.22. Photon correlation measurements of fluorescent light from sodium. The laser is tuned on resonance. The smooth curve was obtained by a fit taking account of the number fluctuations of the atomic beam. The corresponding parameters are compiled in Table 3.4

for $\tau < 0$. This provides an additional test for the data since the second-order correlation function is symmetric around $\tau = 0$. The photomultipliers were cooled to give dark count rates of less than 10 cps. The on-resonance count rates of each multiplier was typically 5000 cps, and off-resonance 500 cps.

Figure 3.22 shows three experimental histograms for three different laser intensities with the laser frequency in resonance with the atomic transition frequency. The solid curves result from least-squares fits of the theoretical function

$$g^{(2)}(\tau) = 1 + \frac{1}{\bar{N}(1 + \frac{\delta}{\bar{N}})^2}\left[1 - e^{-3\Gamma\tau/4}\left(\cos\Omega\tau + \frac{3\Gamma}{4\Omega}\sin\Omega\tau\right)\right] \tag{3.5}$$

to the experimental data. Here $\Omega = \sqrt{\omega^2 - (\Gamma/4)^2}$, Ω is the Rabi frequency, Γ the natural decay rate, $\bar{N}$ the average number of atoms in the observation volume, and $\bar{N}/\delta$ is the signal-to-noise ratio. The function $g^{(2)}(\tau)$ is taken from CARMICHAEL et al. [3.36]. (The number of observed coherence areas is approximately 10^3. The formula is only valid if a large number of coherence areas are observed.)

Note that within the statistics of the experimental data the curves are symmetric around $\tau = 0$. Table 3.4 gives the experimental parameters and those determined by the least-squares fit. The spatial diameter of the laser beam has not been determined accurately enough, therefore, only the laser power is listed.

For higher laser intensities ($\omega \gg \Gamma/4$) the Rabi frequency varies with the square root of the laser power. This is shown in Fig.3.23. The slope of the straight line in this double logarithmic plot is 0.53, in good agreement with the theoretical value of 0.5.

Table 3.4

Experimental data		Parameters obtained by the fit		
P_{Laser} [mW]	Signal to noise $\bar{N}/\delta$	$\bar{N}$	$\tau = \Gamma^{-1}$ [ns][a]	$\Omega/2\pi$ [MHz]
2.5	12.2	5 (1)	12.4 (2.4)	27.2 (1.1)
5	6.0	5 (1)	16.3 (3.7)	37.5 (1.3)
10	6.0	3 (1)	13.5 (2.3)	57.5 (1.2)

[a]The lifetime value in the literature is τ = 16.4 ns

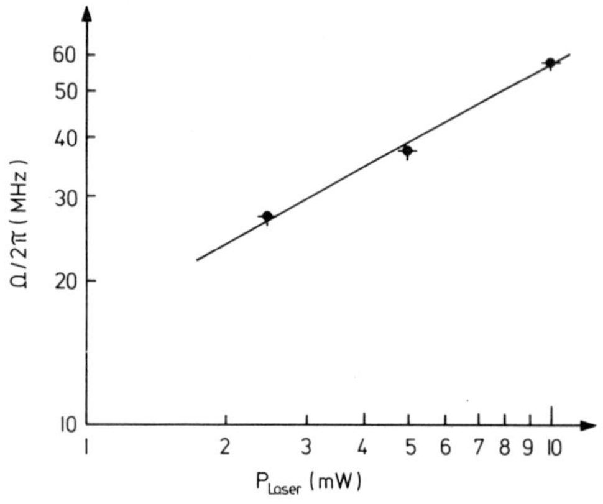

Fig.3.23. Dependence of the Rabi frequency on the laser power for resonant excitation. The slope of the straight line is in very good agreement with the theoretical value of 0.5

The measurement for 10 mW (Fig.3.22) shows the Rabi oscillation to level out for longer times to a value larger than the minimum at τ = 0 which clearly exhibits the antibunching at τ = 0. This is in contrast to the experiments performed by KIMBLE et al. [3.35], where the number of coincidences gets smaller rather fast for large delay values due to the short transit time ($\approx$100 ns) of the atoms through the observation volume. In the present experiment such a decrease has not been observed since the transit time was 250 ns.

By introducing an offset to the modulated magnetic field applied to the reference atomic beam the laser frequency could be stabilized off resonance. Figures 3.24 and 25 show experimental curves for different laser intensities and for the laser frequency on and off resonance. The detuning was $\Delta/2\pi$ = 13.7 MHz and $\Delta/2\pi$ = 17 MHz, respectively. The change of the Rabi frequency due to the detuning can be seen in both figures. The Rabi frequency for nonresonant excitation is given by

$$\Omega' = \sqrt{\Omega^2 + \Delta^2}$$

where Ω is the Rabi frequency for resonant excitation and Δ the detuning. The values for Ω and Ω' deduced from the measurements in Figs.3.24,25, respectively, are given in Table 3.5 together with the calculated value Ω'_{theor}. There is a good agreement between experimental and theoretical values.

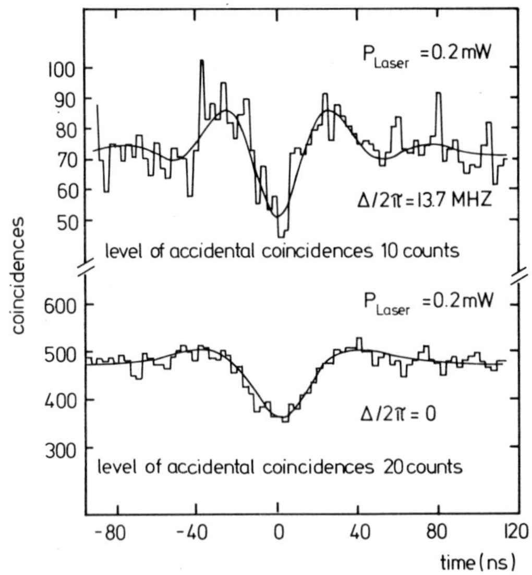

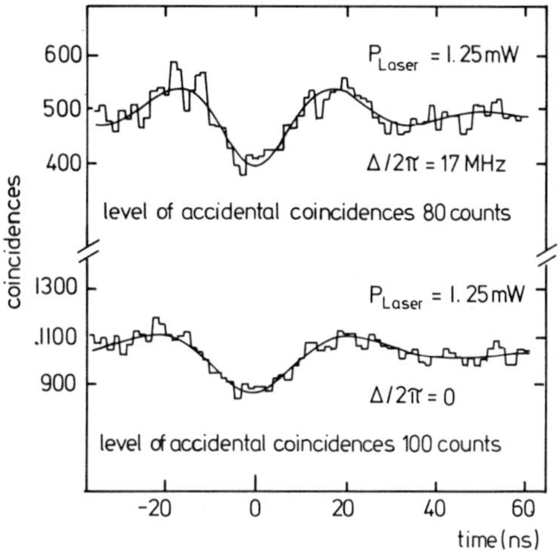

Fig.3.25 ▲

◄ Fig.3.24

Fig.3.24. Photon correlation measurements of fluorescent light from sodium atoms. For the upper measurement the laser was tuned to a frequency differing by $\Delta/2\pi$ = 13.7 MHz from the resonance. The smooth curve was obtained by a fit taking account of the influence of atomic number fluctuations of the atomic beam. The corresponding parameters are compiled in Table 3.5

Fig.3.25. Photon correlation measurements of fluorescent light from sodium atoms. For the upper measurement the laser was tuned to a frequency differing by $\Delta/2\pi$ = 17 MHz from the resonance. The smooth curve was obtained by a fit taking account of the influence of atomic number fluctuations of the atomic beam. The corresponding parameters are compiled in Table 3.5

Table 3.5

Experimental data		Parameters obtained by the fit		Calculated parameters
P_{Laser} [mW]	$\Delta/2\pi$ [MHz]	$\Omega/2\pi$ [MHz]	$\Omega'/2\pi$ [MHz]	$\Omega'_{theor.}/2\pi$ [MHz]
1.25	0	23.4 (0.8)		
1.25	17 (2)		29.5 (0.9)	28.9 (1.8)
0.2	0	12.5 (0.7)		
0.2	13.7 (0.7)		19.4 (1.1)	18.6 (1.0)

According to (3.5) it is obvious that the Rabi oscillations in $g^{(2)}(\tau)$ are washed out with an increasing average number $\bar{N}$ of atoms. In the experiment $\bar{N}$ has been changed systematically. The influence on $g^{(2)}(\tau)$ is shown in Fig.3.26.

The initial results obtained by KIMBLE et al. [3.80] showed for $g^{(2)}(\tau)$ a positive slope characteristic of photon antibunching but starting with $g^{(2)}(0) = 1$ rather than zero. The reason for this was, as discussed above, pointed out by JAKEMAN et al. [3.34] who showed that this is due to the number fluctuations in

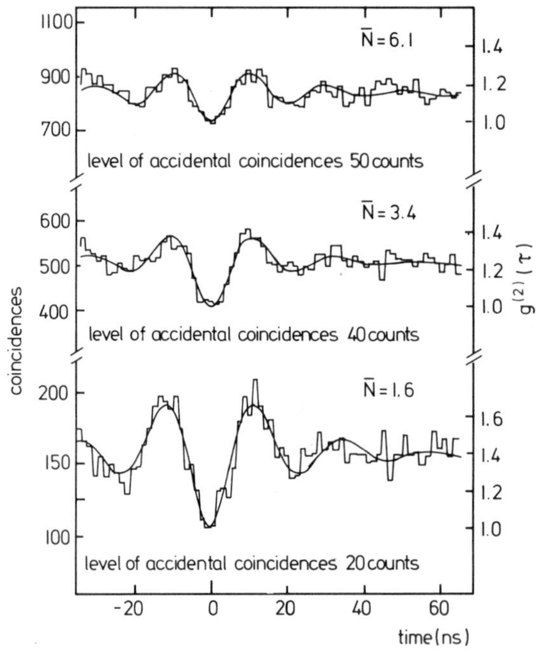

Fig.3.26. Influence of the average atomic number $\bar{N}$ on the intensity correlation function. The oscillations are washed out with increasing $\bar{N}$

the atomic beam. Later KIMBLE et al. [3.35] corrected the measurements for the multiatom and transit-time effects (see also [3.36]). Some recent experimental results by DAGENAIS and MANDEL are shown in Fig.3.27. They also agree quite well with theory. These experiments clearly show the evidence of photon antibunching and thus verify the predictions of the quantum theory of light.

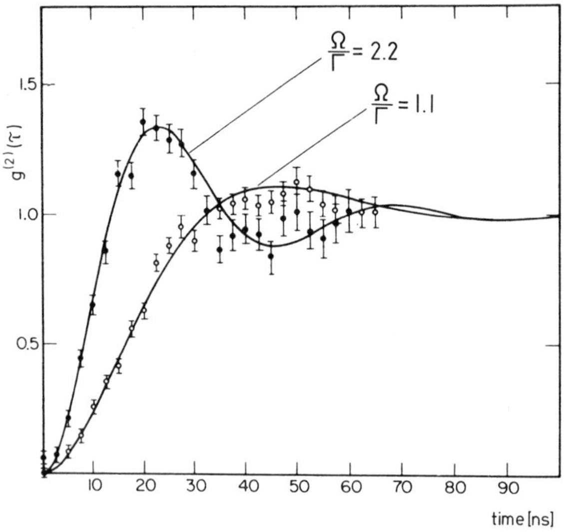

Fig.3.27. Photon correlation measurements in comparison with theory (solid line). The solid points correspond to $\Omega/\Gamma = 2.2$ and the circles to $\Omega/\Gamma = 1.1$ [3.37]

There is still interest in the photon correlation at low laser intensities [3.44] where $\Omega/\Gamma < 1$. In this limit the laser bandwidth changes the photon correlation in a different way compared to the case $\Omega/\Gamma > 1$. The result for low laser intensity gives a generalization of the Heitler-Weisskopf effect (Sect.3.3.1) applied to photon correlations. The signal is described by

$$g^{(2)}(\tau) = 1 + e^{-\tau\Gamma}\,\frac{1 + 2\delta/\Gamma}{1 - 2\delta/\Gamma} - 2\,\frac{e^{-\tau\Gamma/2-\delta\tau}}{1-2\delta/\Gamma}$$

where δ is the diffusion coefficient of the phase of the laser [3.44], i.e., the laser linewidth. For the limit $\delta = 0$ (monochromatic source) it follows that

$$g^{(2)}(\tau) = (1 - e^{-\tau\Gamma/2})^2 \quad ;$$

this is in reasonable agreement with the measurement shown in Fig.3.28 when in addition the finite transit time of the atoms through the observation region is considered.

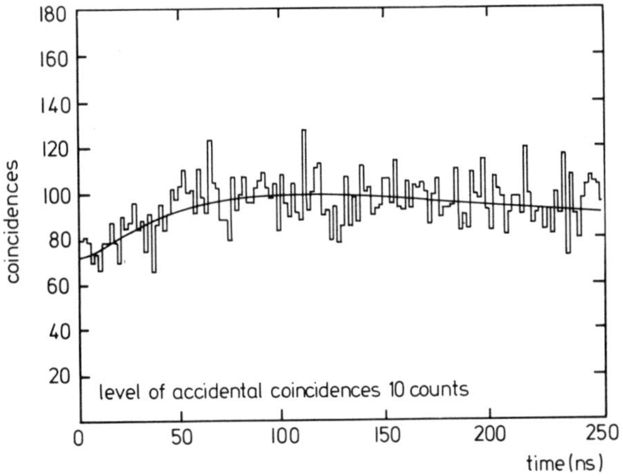

Fig.3.28. Photon correlation for low laser intensities $\Omega/\Gamma < 1$. The theoretical curve (solid line) is corrected for finite transition time effects

A new type of correlation experiment has been performed quite recently by COHEN-TANNOUDJI and coworkers [3.81]. In contrast to the experiments on resonance fluorescence described in Sect.3.3.1, emphasizing either the frequency or the time features, the new experiment deals with a mixed analysis. It investigates the time correlation between fluorescence photons selected by frequency filters. If the three components of the fluorescence triplet are well separated one can use filters centered at any one of these components. Then it is possible to study the time correlation of the filtered fluorescence.

In the experiment a Sr atomic beam is excited by a laser line which is 28 Å off resonance. It was seen that the photons of the two sidebands of the fluorescence triplet are emitted in a well-defined time order which can be explained in terms of the sequence of fluorescence decays down the energy diagram of a dressed atom. This experiment gives, despite the fact that the principal features of the resonance fluorescence in a strong monochromatic laser field are understood, an interesting new view of the processes involved.

Acknowledgement. A part of the experimental work described here was supported by the Deutsche Forschungsgemeinschaft. This support is gratefully acknowledged.

References

3.1 B.R. Mollow: Phys. Rev. *188*, 1969 (1969)
3.2 G. Oliver, E. Ressayre, A. Tallet: Lett. Nuovo Cimento *2*, 77 (1971)
3.3 B.R. Mollow: Phys. Rev. *165*, 145 (1975)
3.4 S.S. Hassan, R.K. Bullough: J. Phys. B*8*, L147 (1975)
3.5 S. Swain: J. Phys. B*8*, L437 (1975)
3.6 C. Cohen-Tannoudji: In *Laser Spectroscopy*, Proc. 2nd Int. Conf., Megève, France, 1975, ed. by S. Haroche, J.C. Pebay-Peyroula, T.W. Hänsch, S.E. Harris, Lecture Notes in Physics, Vol.43 (Springer, Berlin, Heidelberg, New York 1975) p.324
3.7 C. Cohen-Tannoudji: "Atoms in Strong Resonant Fields", in *Frontiers in Laser Spectroscopy*, ed. by R. Balian, S. Haroche, S. Liberman (North-Holland, Amsterdam 1977) Vol.1, p.3
3.8 H.J. Kimble, L. Mandel: Phys. Rev. A*13*, 2123 (1976)
3.9 K. Wodkiewicz, J.H. Eberly: Ann. Phys. N.Y. *101*, 514 (1976)
3.10 H.J. Carmichael, D.F. Walls: J. Phys. B*8*, L77 (1975)
H.J. Carmichael, D.F. Walls: J. Phys. B*9*, L43 (1976)
H.J. Carmichael, D.F. Walls: J. Phys. B*9*, 1199 (1976)
3.11 R.J. Ballagh: Ph. D. Thesis, Univ. of Colorado, USA (1978)
3.12 J.D. Cresser: Ph. D. Thesis, Univ. of Quessland, Australia (1979)
3.13 F. Schuda, C.R. Stroud, Jr., M. Hercher: J. Phys. B*1*, L198 (1974)
3.14 H. Walther: In *Laser Spectroscopy*, Proc. 2nd Int. Conf., Megève, France, 1975, ed. by S. Haroche, J.C. Pebay-Peyroula, T.W. Hänsch, S.E. Harris, Lecture Notes in Physics, Vol.43 (Springer, Berlin, Heidelberg, New York 1975) p.358
3.15 F.Y. Wu, R.E. Grove, S. Ezekiel: Phys. Rev. Lett. *35*, 1426 (1975)
3.16 W. Hartig, W. Rasmussen, R. Schieder, H. Walther: Z. Phys. A*278*, 205 (1976)
3.17 R.E. Grove, F.Y. Wu, S. Ezekiel: Phys. Rev. A*15*, 227 (1977)
3.18 H. Carmichael, D.F. Walls: J. Phys. B*10*, L685 (1977)
3.19 W. Heitler: *Quantum Theory of Radiation*, 3rd ed. (Oxford Univ. Press, London 1964)
3.20 P.A. Apanasevich: Opt. Spectrosc. *16*, 387 (1964)
3.21 P.A. Apanasevich: Opt. Spectrosc. *14*, 324 (1963)
3.22 M.C. Newstein: Phys. Rev. *167*, 89 (1968)
3.23 M. Lax: Phys. Rev. *129*, 2342 (1963)
3.24 C.R. Stroud, Jr.: Phys. Rev. A*3*, 1044 (1977)
3.25 M.E. Smithers, H.S. Freedhoff: J. Phys. B*8*, L209 (1975)
3.26 B. Renaud, R.M. Whitley, C.R. Stroud, Jr.: J. Phys. B*9*, L19 (1976)
3.27 B. Renaud, R.M. Whitley, C.R. Stroud, Jr.: J. Phys. B*10*, 19 (1977)
3.28 C.R. Stroud, Jr., E.J. Jaynes: Phys. Rev. A*1*, 106 (1970)
3.29 J.R. Ackerhalt: Phys. Rev. A*17*, 471 (1978)
3.30 M. Lax: Phys. Rev. *157*, 213 (1967)
3.31 K. Wodkiewicz: Phys. Lett. A*73*, 94 (1979)
3.32 S. Swain: J. Phys. A*8*, 1277 (1975)
3.33 J.D. Cresser, B.J. Dalton: J. Phys. A*13*, 795 (1980)
3.34 E. Jakeman, E.R. Pike, P.N. Pusey, J.M. Vaughan: J. Phys. A*10*, L257 (1977)
3.35 H.J. Kimble, M. Dagenais, L. Mandel: Phys. Rev. A*18*, 201 (1978)
3.36 H.J. Carmichael, P. Drummond, P. Meystre, D.F. Walls: J. Phys. A*11*, L121 (1978)

3.37 M. Dagenais, L. Mandel: Phys. Rev. A*18*, 2217 (1978)
3.38 H.D. Simeon, R. London: J. Phys. A*8*, 539 (1975)
3.39 K.J. M'Neil, D.F. Walls: J. Phys. A*7*, 617 (1974)
3.40 D. Stoler: Phys. Rev. Lett. *33*, 1397 (1974)
3.41 H. Paul, W. Brunner: Opt. Acta *27*, 263 (1980)
3.42 A. Bandilla, H.-H. Ritze: Opt. Commun. *28*, 126 (1979)
3.43 H.-H. Ritze, A. Bandilla: Opt. Commun. *28*, 241 (1979)
3.44 K. Wodkiewicz: Phys. Lett. A*77*, 315 (1980)
3.45 M. Schubert, K.-E. Süsse, W. Vogel: Opt. Commun. *30*, 275 (1979)
3.46 M. Schubert, K.-E. Süsse, W. Vogel, D.-G. Welch: Opt. Quantum Electron. *12*, 65 (1980)
3.47 H.J. Kimble, M. Dagenais, L. Mandel: Phys. Rev. Lett. *39*, 691 (1977)
3.48 D.F. Walls: Nature London *280*, 451 (1979)
3.49 G.S. Agarwal: Phys. Rev. A*15*, 2380 (1977)
3.50 J.H. Eberly: Phys. Rev. Lett. *37*, 1381 (1976)
3.51 P. Avan, C. Cohen-Tannoudji: J. Phys. B*10*, 155 (1977)
3.52 H.J. Kimble, L. Mandel: Phys. Rev. A*15*, 689 (1977)
3.53 P. Zoller: J. Phys. B*10*, L321 (1977)
3.54 P.L. Knight, W.A. Molander, C.R. Stroud, Jr.: Phys. Rev. A*17*, 1547 (1978)
3.55 B. Sobolewska: Opt. Commun. *19*, 185 (1976)
3.56 B. Sobolewska, R. Sobolewski: Opt. Commun. *26*, 211 (1978)
3.57 R. Kornblith, J.H. Eberly: J. Phys. B*11*, 1545 (1978)
3.58 P. Avan, C. Cohen-Tannoudji: J. Phys. Paris *36*, L85 (1975)
3.59 C. Cohen-Tannoudji, S. Reynaud: J. Phys. B*10*, 345 (1977)
3.60 G.S. Agarwal: Phys. Rev. Lett. *37*, 1383 (1976)
3.61 P. Zoller, F. Ehlotzky: J. Phys. B*10*, 3023 (1977)
3.62 M.G. Raymer, J. Cooper: Phys. Rev. A*20*, 2238 (1979)
3.63 A.T. Georges, P. Lambropoulos, P. Zoller: Phys. Rev. Lett. *42*, 1609 (1979)
3.64 A.T. Georges: Phys. Rev. A*21*, 2034 (1980)
3.65 M. Le Berre-Rousseau, E. Ressayre, A. Tallet: Phys. Rev. *22*, 240 (1980)
3.66 A.T. Georges, S.N. Dixit: Phys. Rev. (to be published)
3.67 H. Haken: "Laser Theory", in *Light and Matter Ic*, ed. by L. Genzel, Handbuch der Physik, Vol. XXV/2c (Springer, Berlin, Heidelberg, New York 1970) p.130
3.68 C. Cohen-Tannoudji, S. Reynaud: J. Phys. B*10*, 365 (1977)
3.69 H.M. Gibbs, T.N.C. Venkatesan: Opt. Commun. *17*, 87 (1976)
3.70 R. Schieder, H. Walther: Z. Phys. *270*, 55 (1974)
3.71 H. Kogelnik, E.R. Ippen, A. Dienes, Ch.V. Shank: IEEE J. Quantum Electron. *QE-8*, 373 (1972)
3.72 H. Walther: Phys. Scr. *9*, 297 (1974)
3.73 H. Walther: "Atomic and Molecular Spectroscopy with Lasers", in *Laser Spectroscopy of Atoms and Molecules*, ed. by H. Walther, Topics in Applied Physics, Vol.2 (Springer, Berlin, Heidelberg, New York 1976) pp.1-124
3.74 W. Rasmussen, R. Schieder, H. Walther: Opt. Commun. *12*, 315 (1974)
3.75 J. Häger: Ph. D. Thesis, Universität Köln (1975)
3.76 D.A. Jackson, Duong Hong Tuan: Proc. R. Soc. London A*280*, 323 (1964)
3.77 D.A. Jackson, Duong Hong Tuan: Proc. R. Soc. London A*291*, 9 (1966)
3.78 R.J. Glauber: Phys. Rev. *130*, 2529 (1963); *131*, 2766 (1963)
3.79 R. Hanbury-Brown, R.Q. Twiss: Nature Lodon *177*, 27 (1956); Proc. R. Soc. London A*242*, 300 (1957); A*243*, 291 (1957)
3.80 H.J. Kimble, M. Dagenais, L. Mandel: Phys. Rev. Lett. *39*, 691 (1977)
3.81 A. Aspect, G. Roger, S. Reynaud, J. Dalibard, C. Cohen-Tannoudji: Phys. Rev. Lett. *45*, 617 (1980)

4. Theory of Optical Bistability

R. Bonifacio and L. A. Lugiato

With 12 Figures

We present in extenso the theory of optical bistability in a ring cavity. The plane-wave approximation is used throughout. We illustrate first the semiclassical and second the quantum-statistical treatment.

The first is based on the Maxwell-Bloch equations that are solved exactly at steady state both in the purely absorptive and in the mixed absorptive + dispersive case.

We show that in the double limit $\alpha_{abs}L \to 0$, $T \to 0$, with $C = \alpha_{abs}L/2T$ constant, the exact solution reduces to the state equation of the so-called mean-field theory of optical bistability. This theory is then used to discuss the main features of the transient behavior in purely absorptive optical bistability, as the critical slowing down, the Rabi oscillations in the "bad cavity" case, etc. Again on the basis of the Maxwell-Bloch equations, we show that under suitable conditions a part of the high-transmission branch of the hysteresis cycle of transmitted vs incident light becomes unstable. In this case, the system can exhibit a "self-pulsing" behavior, in which the transmitted light consists of an undamped periodic sequence of short pulses. Thus, the system works as a converter of CW light into pulsed light. The quantum-mechanical treatment is based on a suitable one-mode master equation which is the quantum statistical version of the mean-field model. We discuss the photon statistics and the sectrum of transmitted light in purely absorptive optical bistability, for a fluctuationless incident field.

Photon statistics is described by a distribution function which is generally two peaked. The behavior of the mean value and of the fluctuations of the transmitted light supports the analogy between optical bistability and first-order phase transitions in equilibrium systems. On the other hand, the nonthermodynamic character of the transition in optical bistability appears manifest from the results. The spectrum of the transmitted light exhibits a dramatic hysteresis cycle with line narrowing at the boundaries of the cycle. In the "bad cavity" case, one has the discontinuous appearence of a triplet spectrum when the incident field intensity is increased (discontinuous dynamical Stark effect).

4.1 Background

Optical bistability, the topic we shall discuss in this article, is the name of a phenomenon which arises in the transmission of light by an optical cavity filled with a resonant medium. In fact, let a CW coherent beam of intensity I_I be injected into a resonant cavity (e.g., a Fabry-Perot) tuned or nearly tuned to the incident light (Fig.4.1). The injected beam is partially transmitted, partially reflected, and partially scattered by the cavity; let I_T and I_R be the transmitted and reflected intensities, respectively. When the cavity is empty one has $I_T = aI_I$ where the proportionality constant a depends on cavity mistuning and on the transmissivity coefficient T of the mirrors, in particular, a = 1 for perfect tuning. On the other hand, when the cavity is filled with material resonant or nearly resonant with the incident field, a is nonlinear function of I_I. The shape of this function crucially depends on the parameter $C = \alpha_{abs}L/2T$, where α_{abs} is the linear absorption coefficient of the material and L is the length of the cavity. If $C \ll 1$ one has practically the empty-cavity situation. When C is of the order of unity I_T is still a continuous and monotonic function of I_I, but there is a portion of the curve I_T vs I_I in which the differential gain dI_T/dI_I is larger than unity (Fig.4.2a). In such conditions the system behaves as an *optical transistor*. In fact, if the incident intensity is adiabatically modulated around a value of I_I such that $dI_T/dI_T > 1$ the modulation gets amplified in the transmitted light. This amplification effect increases with C. Finally, if C exceeds a suitable critical value which depends both on the cavity mistuning and on the atomic detuning, the transmitted intensity varies discontinuously with the incident light showing a hysteresis cycle (Fig.4.2b). In this cycle there is a low-transmission branch and a high-transmission branch. Hence there is a suitable range of values of I_I such that the system has two different stationary states; this is *optical bistability* (OB). For zero atomic detuning δ_A one has *absorptive OB or bistable absorption*, which arises from the bleaching effect of the absorber for high incident intensity. On the other hand, when $\delta_A \neq 0$ dispersion also plays a role, and for large enough δ_A one has purely *dispersive OB* which arises from the interplay of atomic detuning and cavity mistuning. Of course, both absorptive and dispersive OB can be studied in homogeneously as well as inhomogeneously broadened atomic systems.

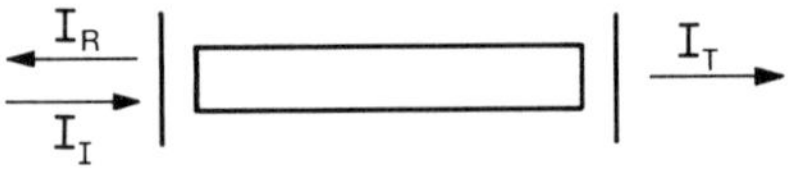

Fig.4.1. Resonant cavity, I_I, I_T, and I_R are the incident, transmitted, and reflected intensities, respectively

Absorptive OB has been theoretically predicted by SZÖKE et al. [4.1]. Some years later McCALL [4.2] predicted the transistor effect and treated absorptive OB in a Fabry-Perot cavity by a numerical analysis of two-sided Maxwell-Bloch equations.

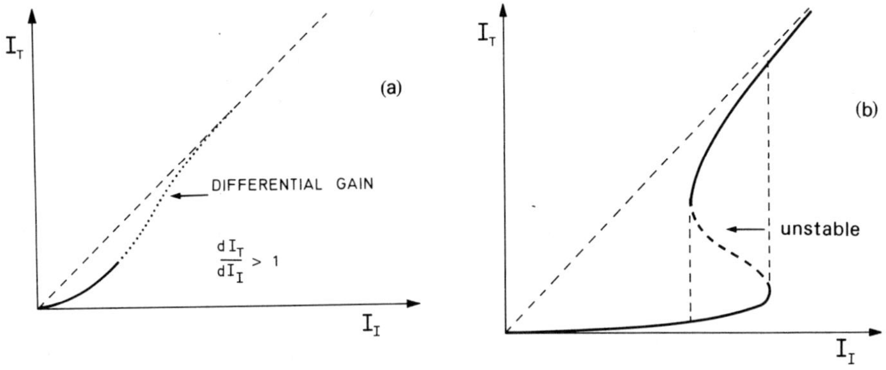

Fig.4.2a,b. Nonlinear response of a filled cavity for zero cavity mistuning, (a) transistor operation, (b) optical bistability

This work suggested the experiments of GIBBS, McCALL, and VENKATESAN in Na and Ruby, in which both transistor operation and bistability were observed [4.3,4]. The analysis of the data showed that the observed bistability was mainly of dispersive type. The mechanism which produces dispersive OB was explained with the help of a simple phenomenological "cubic" model. These results stimulated theoretical and experimental activity; see especially FELBER and MARBURGER [4.5], BONIFACIO and LUGIATO [4.6], SMITH and TURNER [4.7]. Subsequently, experiments on OB have been performed by BISHOFSBERGER and SHEN [4.8], GRISCHKOWSKI [4.9], GARMIRE et al. [4.10], GIBBS et al. [4.11], MILLER et al. [4.12], SANDLE and GALLAGHER [4.13], GRYNBERG et al. [4.14], ARIMONDO et al. [4.15].

A complete and analytical theory was first formulated in the case of absorptive OB in a ring cavity, perfectly tuned to the incident field. In fact, in this situation one can calculate analytically and exactly the stationary states [4.16] and explicitly analyse the stability of these solutions [4.17]. This analysis leads to the prediction that under suitable conditions (explicitly specified by the treatment) the system exhibits self-pulsing behavior [4.18]. This possibility suggests the use of this device as a *converter of CW light into pulsed light*. The exact analytical treatment of the steady state given in [4.16] has been extended to the mixed absorptive-dispersive case independently in [4.19-21], while in [4.22] one has the generalization for Lorentzian inhomogeneous broadening. Furthermore the results of [4.16] with the analysis of the mean-field limit (4.11) have been generalized for the Fabry-Perot cavity, numerically [4.23-25] and analytically [4.26-28], in both the absorptive [4.23-27] and the mixed absorptive-dispersive [4.24,28] cases.

Furthermore, the analysis in [4.16] has given a fundamental justification of the previously formulated mean-field theory (MFT) of OB [4.6,29,30], in which the propagation of the electric field is treated in an average way. This theory allows obtaining a simple description of OB which gives physical insight and sheds light on many fascinating aspects of this phenomenon. In particular, the semiclassical treatment of the steady state clarifies the role of atomic cooperation in OB and reveals the deep connections between optical bistability, superfluorescence, and resonance fluorescence. New predictions have been given for the transient [4.29-34], which are relevant for the discussion of the switching properties as well as for the description of the spectrum of the transmitted light.

A relevant advantage of the mean-field model is that it has a natural quantum-mechanical formulation, thereby allowing one to deal with the fluctuations of the system. The quantum-mechanical analysis of the stationary situation points out the striking similarity of OB with first-order phase transitions in equilibrium systems, showing on the other hand the nonthermodynamic character of the transition [4.35-41]. The spectrum of transmitted light is shown to undergo a hysteresis cycle, with *line narrowing* in correspondence to the discontinuity points of the cycle; under suitable conditions, one has a sudden transition from a single narrow line to a triplet with well-separated sidebands (*discontinuous dynamical Stark effect*) [4.30,35,42-46].

The mean-field model for a ring cavity has been generalized, both at the semiclassical and at the quantum-statistical level, for mixed absorptive-dispersive bistability [4.47]. The general bistability conditions have been worked out in [4.48-50]. In particular, in [4.50] we analyzed in detail the behavior of the hysteresis cycle when the parameters in play are varied, and showed that one finds the largest cycle in the purely absorptive case. The limits of the validity of the "cubic" model of [4.3] are discussed in the case of two-level atoms. Further analyses of dispersive bistability can be found in [4.51-54].

OB in two-photon transitions was discussed theoretically in [4.55] (absorptive case) and [4.56] (dispersive case) and was observed experimentally in [4.14].

In this chapter we shall deal exclusively with the theory of OB in a ring cavity. Some details of the treatment of OB in a Fabry-Perot, together with a discussion of some relevant experiments, are given in Chap.5. Further material and references on the general subject of OB can be found in [4.57].

4.2 Theory of Absorptive OB in a Ring Cavity

To formulate a theory of OB it is suitable to consider a ring cavity because one avoids the difficulties connected with standing waves. As shown by Fig.4.3, we consider a sample of length L and volume V containing $N \gg 1$ two-level atoms and placed in a ring cavity of total length $\mathcal{L} = 2(L + \ell)$. E_I is the incident field amplitude. E_T and E_R are the transmitted and reflected field amplitudes, respectively. The upper mirrors have reflectivity coefficient $R = 1 - T$ whereas the lower mirrors have 100% reflectivity. We assume that

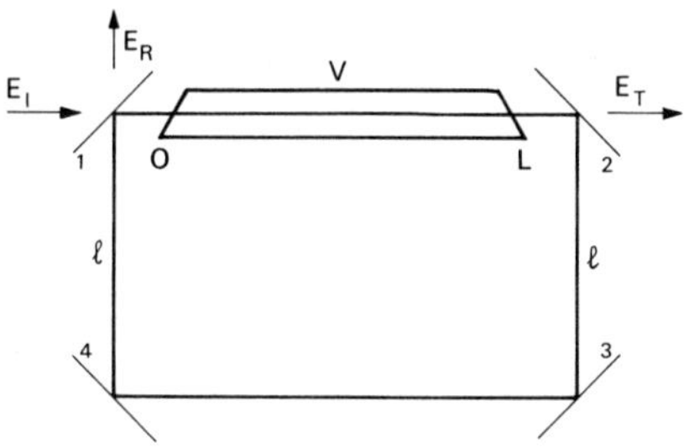

Fig.4.3. Ring cavity, E_I, E_T, and E_R are the incident, transmitted, and reflected amplitudes, respectively. Mirrors 1 and 2 have reflectivity coefficient $R = 1-T$, while mirrors 3 and 4 have 100% reflectivity

1) the atomic system is homogeneously broadened and the transition frequency of the atoms coincides with the frequency ω_0 of the incident field (zero atomic detuning), and

2) L is equal to an integer number of wavelengths (zero cavity mistuning).

4.2.1 Semiclassical Theory

a) *Exact Treatment of the Stationary Situation*

The incident field propagates in the sample, while on the other hand the atomic system is driven by the incident field and reacts on it. These dynamics are described by the well-known Maxwell-Bloch equations with phenomenological damping terms:

$$\frac{\partial P}{\partial t} = \frac{\mu}{\hbar} E\Delta - \gamma_\perp P \quad , \tag{4.1a}$$

$$\frac{\partial \Delta}{\partial t} = -\frac{\mu}{\hbar} EP - \gamma_\parallel \left(\Delta - \frac{N}{2}\right) \quad , \tag{4.1b}$$

$$\frac{\partial E}{\partial t} + c\frac{\partial E}{\partial z} = -gP \quad , \tag{4.1c}$$

where P is the macroscopic polarization field, Δ is one-half the difference between the total populations of the lower and of the upper level, E is the slowly varying envelope of the electric field, μ is the modulus of the dipole moment of the atoms, g is the coupling constant

$$g = \frac{4\pi\omega_0}{V}\mu \quad , \tag{4.2}$$

and $\gamma_\parallel = T_1^{-1}$ and $\gamma_\perp = T_2^{-1}$ are the longitudinal and transverse atomic relaxation rates. The boundary conditions for this problem are

$$E(L,t) = E_T(t)/\sqrt{T} \quad , \quad E(0,t) = \sqrt{T}E_I + RE(L,t-\Delta t) \quad , \tag{4.3a}$$

$$E_R(t) = \sqrt{RT}E(L,t-\Delta t) - \sqrt{R}E_I = \sqrt{R}[E_T(t-\Delta t) - E_I] \tag{4.3b}$$

where $\Delta t = (L + 2\ell)/c$ is the time the light takes to travel from mirror 2 to mirror 1 (Fig.4.3). E_I is assumed real and positive for definiteness.

Let us consider the stationary situation ($\partial P/\partial t = \partial\Delta/\partial t = \partial E/\partial t = 0$). When the cavity is empty one obtains $E(L) = E(0)$, i.e., by (4.3a,b) $E_T = E_I$. For a filled cavity we obtain from (4.1a,b)

$$P(z) = \frac{N}{2}\sqrt{\frac{\gamma_\parallel}{\gamma_\perp}}\frac{X(z)}{1 + X^2(z)} , \tag{4.4a}$$

$$\Delta(z) = \frac{N}{2}\frac{1}{1 + X^2(z)} , \tag{4.4b}$$

where $X(z)$ is the normalized field amplitude

$$X(z) = \frac{\mu E(z)}{\hbar\sqrt{\gamma_\perp\gamma_\parallel}} . \tag{4.4c}$$

Substituting (4.4a) into (4.1c) and using (4.4c) we obtain the stationary field equation

$$\frac{dX}{dz} = -\alpha_{abs}\frac{X}{1 + X^2} , \tag{4.5}$$

where α_{abs} is the linear absorption coefficient per unit length

$$\alpha_{abs} = \frac{\mu g N}{2\hbar c\gamma_\perp} . \tag{4.6}$$

Let us now define normalized incident and transmitted fields y and x as follows:

$$y = \mu E_I/\hbar\sqrt{\gamma_\perp\gamma_\parallel T} ,$$

$$x = X(L) = \mu E_T/\hbar\sqrt{\gamma_\perp\gamma_\parallel T} . \tag{4.7}$$

With these notations the boundary condition (4.3a) can be rewritten as

$$X(0) = Ty + Rx . \tag{4.8}$$

Equation (~~4.4~~ 4.5) can be immediately integrated obtaining

$$\ln[X(0)/x] + \frac{1}{2}[X^2(0) - x^2] = \alpha_{abs}L . \tag{4.9}$$

Finally by combining (4.8,9) we get the exact relation between the transmitted field x and the incident field y,

$$\ln\left[1 + T\left(\frac{y}{x} - 1\right)\right] - \frac{x^2}{2}\left\{\left[1 + T\left(\frac{y}{x} - 1\right)\right]^2 - 1\right\} = \alpha_{abs}L . \tag{4.10}$$

Relation (~~4.12~~ 4.10) only depends on two parameters, $\alpha_{abs}L$ and T. Let us now consider the following double limit:

$$T \to 0 \quad , \quad \alpha_{abs}L \to 0 \quad , \quad \frac{\alpha_{abs}L}{2T} = C \text{ constant} \quad . \tag{4.11}$$

In this limit $\varepsilon \overset{def}{=} T(y/x - 1)$ is infinitesimal, so that (4.10) reduces to $\varepsilon(1 + x^2) = \alpha_{abs}L$, which can be rewritten as

$$y = x + \frac{2Cx}{1 + x^2} \quad . \tag{4.12}$$

Hence we have the following *theorem*: in the limit (4.11) one obtains the *state equation* (4.12) of the mean-field theory of OB in a ring cavity [4.6] which depends on the single parameters C. In fact, in the MFT the space variation of the fields is neglected, so that (4.5) is integrated as follows:

$$X(L) - X(0) = -\alpha_{abs} \int_0^L dz \frac{X(z)}{1 + X^2(z)}$$

$$\simeq -\alpha_{abs} L \frac{X(L)}{1 + X^2(L)} \quad , \tag{4.13}$$

so that using the boundary condition (4.8) and (4.7) one obtains (4.12). This theorem can be intuitively understood on the basis of the fact that for small $\alpha_{abs}L$ the field varies slowly with z. However, the limit $\alpha_{abs}L \to 0$ alone gives trivially the empty cavity solution, which corresponds to C = 0. Only operating the double limit (4.11) preserves the arbitrariness of the parameter C. Figure 4.4 shows how the limit (4.12) is approached in the case C = 10. Clearly the mean-field result (4.12) is very good up to $\alpha_{abs}L \simeq 2$; this is the best justification of the MFT. From Fig.4.4 one sees also that, increasing T enough, x becomes a monotonic function of y, so that bistable behavior disappears. In particular, this situation always occurs for T = 1. This result shows a very important point: to achieve bistability one needs not only a saturable absorber, but also good mirrors. In fact, in absorptive OB bistability arises from the combined action of atomic absorption and feedback from the mirrors.

A state equation formally identical to (4.12) was previously derived by SZÖKE et al. [4.1] by phenomenological arguments for a Fabry-Perot. The more precise treatment in [4.25,26,28], which takes standing-wave effects fully into account, shows that in the mean-field limit (4.11) one obtains a different state equation which coincides (apart from a change of sign to convert the absorber into an amplifier) with that previously derived in [4.58] for a lser with injected signal. However the exact state equation for a Fabry-Perot is well approximated (within 15% error) by (4.12) provided that the definition of the quantities x, y, and C is slightly modified as in [Ref.4.30, App.A].

In deriving the state equation (4.12) we have assumed that the field is a plane wave. The effect of the transverse profile of the field in the cavity is discussed in [4.59], both for a ring and for a Fabry-Perot cavity (see also [4.15]).

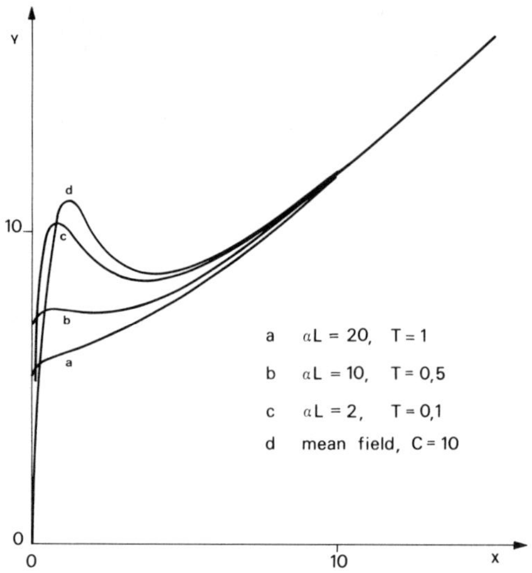

Fig.4.4. Plot of incident light vs transmitted light for $C = \alpha L/2T$ fixed equal to 10 and different values of αL and T. For $\alpha L \to 0$, $T \to 0$ one approaches the behavior predicted by mean-field theory

b) *Mean-Field Approach: Steady-State Analysis*

The semiclassical formulation of the MFT is based on the three equations

$$\dot{P} = \frac{\mu}{\hbar} E\Delta - \gamma_{\perp} P \quad , \tag{4.14a}$$

$$\dot{\Delta} = -\frac{\mu}{\hbar} EP - \gamma_{\parallel}\left(\Delta - \frac{N}{2}\right) \quad , \tag{4.14b}$$

$$E = -g\frac{L}{\mathcal{L}} P - k\left(E - \frac{E_I}{\sqrt{T}}\right) \quad , \tag{4.14c}$$

where P(t), Δ(t), and E(t) are the space averages of P(z,t), Δ(z,t), and E(z,t), respectively;

$$P(t) = \frac{1}{L}\int_0^L P(z,t)dz \quad , \quad \text{etc.};$$

and dots mean derivation with respect to time. k is the cavity halfwidth (or cavity damping constant)

$$k = \frac{cT}{\mathcal{L}} \quad . \tag{4.15}$$

Equations (4.14a-c) follow from (4.1) in the limit of small space variation of the fields [4.30] (this amounts to considering only the mode of the cavity that is resonant with the incident field, neglecting all the off-resonance modes). In fact, if we take the space average of both members of (4.1a,b) and factorize the average of products into the product of averages, we obtain (4.14a,b). On the other hand, by taking the space average of (4.1c) we get

$$\dot{E}(t) = -\frac{c}{L}[E(L,t) - E(0,t)] - gP \quad . \tag{4.16a}$$

Now using the boundary conditions (4.3a) we have for $R \simeq 1$

$$\begin{aligned} E(L,t) - E(0,t) &= E(L,t) - RE(L,t-\Delta t) - \sqrt{T}E_I \\ &\simeq (1-R)E(L,t) + R\Delta t \frac{dE(L,t)}{dt} - \sqrt{T}E_I \\ &\simeq \sqrt{T}(E_T - E_I) + \frac{\Delta t}{\sqrt{T}}\frac{dE_I}{dt} \quad , \end{aligned} \tag{4.16b}$$

where we have truncated the Taylor expansion of $E(L,t-\Delta t)$ to the second term, assuming that Δt is much smaller than all the characteristic times which rule the phenomenon. Finally by taking into account the relations

$$E_T(t) = \sqrt{T}E(L,t) = \sqrt{T}E \quad , \tag{4.16c}$$

one verifies that (4.16a,b) give (4.14c).

The internal field is composed by the incident field $E_I/\sqrt{T}$ and by the reaction field $E - E_I/\sqrt{T}$ which is emitted by the atoms themselves. The reaction field is, as usual, the very cause of atomic cooperation [4.60]. Using (4.7,16) and the relation

$$C = \mu g N / 4\hbar\gamma_\perp k \tag{4.17}$$

one easily verifies that at steady state ($\dot{P} = \dot{\Delta} = \dot{E} = 0$) (4.14a,b) lead to (4.4a,b) with $X(z)$ replaced by x, while (4.14c) gives the state equation (4.12). Let us now analyse this equation which expresses the incident field y as a function of the transmitted field x. The nonlinear term $2Cx/(1 + x^2)$ arises from the reaction field and hence from atomic cooperation, which is "measured" by the parameter C. For very large x (4.12) reduces to the empty-cavity solution $x = y$ (i.e., $E_T = E_I$). The atomic system is saturated so that the medium is bleached. In this situation each atom interacts with the incident field as if the other atoms were not there; this is the noncooperative situation, and in fact the quantum-statistical treatment shows that atom-atom correlations are negligible [4.44]. On the other hand, for small x (4.12) reduces to $y = (2C + 1)x$. Here the linearity arises simply from the fact that for small external fields the response of the system is linear. In this situation the atomic system is unsaturated; for large C the atomic cooperation is dominant and one has strong atom-atom correlations [4.44].

The curves $y(x)$ obtained by varying C are analogous to the van der Waals curves for the liquid-vapour phase transition, with y, x, and C playing the role of pressure, volume and temperature, respectively. For $C < 4$ y is a monotomic function of x, so that one has no bistability (Fig.4.5). However, in part of the curve the differential gain dx/dy is larger than unity, so that in this situation one has the possibility of transistor operation. For $C = 4$ (critical curve) the graph has an

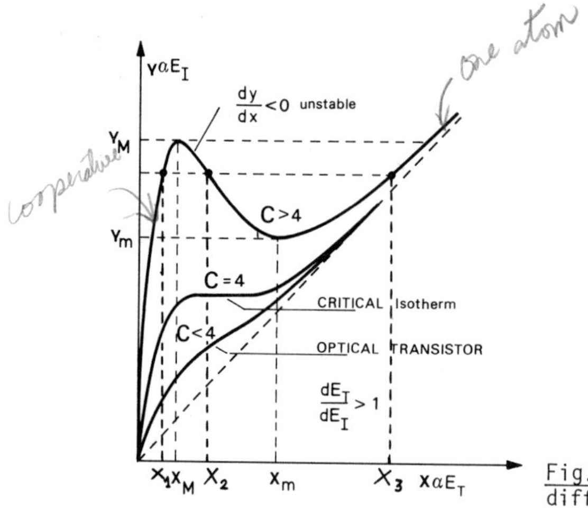

Fig.4.5. Plot of the function $y=x+2Cx/(1+x^2)$ for different values of the parameter C

inflection point with horizontal tangent. Finally for $C > 4$ the curve develops a maximum and a minimum, which for $C \gg 1$ correspond to $(x_M \simeq 1,\ y_M \simeq C)$ and $(x_m \simeq \sqrt{2C},\ y_m \simeq \sqrt{8C})$. Hence for $y_m < y < y_M$ one finds three stationary solutions $x_1 < x_2 < x_3$. As we shall show in Sect.4.2.1d, solutions x_2 on the part of the curve with negative slope are unstable. Therefore we have a bistable situation and by exchanging the axes x and y we immediately obtain the hysteresis cycle of transmitted-vs-incident light (Fig.4.2b). Since atomic cooperation is dominant in the states x_1 and negligible in the states x_3, we shall call x_1 "cooperative stationary state" and x_3 "one-atom stationary state".

c) *Mean-Field Approach: Transient Behavior*

Let us consider a stable stationary state of the system. If the state of the system is slightly shifted from this stationary state, the regression to the steady state is ruled by the system of equations obtained by linearizing (4.14). To be more specific, we can consider the following experiment. Let us assume that the system is initially in a steady state corresponding to some value E_I of the incident field. If E_I is rapidly changed into $E_I + \delta E_I(|\delta E_I| \ll E_I)$ the system approaches the new slightly different steady state corresponding to $E_I + \delta E_I$. This approach is described by a solution of the linearized equations (4.14) and can by experimentally observed by looking at the transient behavior of the transmitted light. The solutions of the linearized equations are linear combinations of three (five, if E and P are taken complex) exponentials $\exp(-\lambda_i t)$, $i = 1,2,3$. When the decay constants λ_i are well separated, the approach to the stationary situation is mainly characterized by the decay constant $\bar{\lambda}$ which has the smallest real part. One proves in general the following results:

1) Let us consider a point (x,y) on the cooperative branch very near to the upper discontinuity point $(x = x_M,\ y = y_M)$. The approach to the steady state (x,y) is

very slow, and is the slower the nearer y is to y_M. Hence there is a *critical slowing down* in correspondence to the discontinuity point y_M. This critical slowing down is similar to that found in tunnel diodes [4.61]. More specifically, one finds that $\bar{\lambda} \to 0$ as $(y_M^2 - y^2)^{1/2}$ when y approaches y_M from below.

2) A similar critical slowing down is found in correspondence to the lower discontinuity point $(x = x_m,\ y = y_m)$. In fact, let us consider a point x,y on the one-atom branch very near to $(x = x_m,\ y = y_m)$. The damping constant $\bar{\lambda}$ which characterizes the approach to x,y is real and tends to zero as $(y^2 - y_m^2)^{1/2}$ when y approaches y_m from above.

The other features of the behavior of $\bar{\lambda}$ when y is varied depend on the relative order of magnitude of the constants k, $\gamma_\perp$, $\gamma_\parallel$. Two typical situations, which we shall consider in the following are:

1) $k \ll \gamma_\perp, \gamma_\parallel$. In this situation, which is usual in laser amplifiers, the empty-cavity width is much smaller than the atomic linewidth. We shall call this the "good quality cavity case".

2) $k \gg \gamma_\perp, \gamma_\parallel$. We shall call this situation, which is typical of superfluorescence, the "bad quality cavity case".

In case 1), $\bar{\lambda}$ is always real, so that the approach to the steady state is always monotonic. One has precisely

$$\bar{\lambda} = k\,\frac{dy}{dx} = k\left[1 + 2C\,\frac{1 - x^2}{(1 + x^2)^2}\right] \quad . \tag{4.18}$$

In case 2), when x,y lies on the cooperative branch, $\bar{\lambda}$ turns out to be real. For $C \gg 1$ one has approximately

$$\bar{\lambda} = 2\gamma_\parallel\left(1 - \frac{y^2}{C^2}\right)^{1/2} \Big/ \left[1 + \left(1 - \frac{y^2}{C^2}\right)^{1/2}\right] \quad . \tag{4.19}$$

When x,y lies on the one-atom branch $\bar{\lambda}$ is complex, except when x,y is very near to the lower discontinuity point $(x_m,\ y_m)$. Hence the approach to the one-atom steady state is oscillatory. In particular, for $y \gtrsim y_M$ one has

$$\bar{\lambda} \simeq (\gamma_\perp + \gamma_\parallel)/2 \pm i\Omega_I \quad , \tag{4.20}$$

where Ω_I is the Rabi frequency of the incident field

$$\Omega_I = \mu E_I/\hbar\sqrt{T} \quad . \tag{4.21}$$

The approach to the steady state when the system is initially far from the stationary situation is analyzed in [4.31,33], both in the framework of the MFT and by numerically solving the MBE. Again it turns out that for T small enough the mean-field approximation works quite satisfactorily.

The predicted critical slowing down in correspondence to the upper bistability threshold has been recently experimentally observed in a hybrid device [4.10].

d) *Complete Linear Stability Analysis*

Let $E_{st}(z)$, $P_{st}(z)$, and $\Delta_{st}(z)$ be one of the stationary solutions of the MBE (4.1) for some given y. To analyse the stability of such a solution we introduce the deviations

$$\delta E(z,t) = E(z,t) - E_{st}(z) \quad , \quad \text{etc.} \tag{4.22}$$

and linearize (4.1) around the stationary solution obtaining

$$\frac{\partial \delta P}{\partial t} = \frac{\mu}{\hbar}(E_{st}\delta\Delta + \Delta_{st}\delta E) - \gamma_{\perp}\delta P \quad , \tag{4.23a}$$

$$\frac{\partial \delta\Delta}{\partial t} = -\frac{\mu}{\hbar}(E_{st}\delta P + P_{st}\delta E) - \gamma_{\|}\delta\Delta \quad , \tag{4.23b}$$

$$\frac{\partial \delta E}{\partial t} + c\,\frac{\partial \delta E}{\partial z} = -\,g\delta P \quad , \tag{4.23c}$$

where δE must obey the boundary condition

$$\delta E(0,t) = R\delta E(L,t - \Delta t) \quad . \tag{4.24}$$

We look for solutions of (4.23) of the form

$$\begin{aligned} \delta E_{\lambda}(z,t) &= \delta E_{\lambda}(z)\exp(\lambda t) + \text{c.c.} \quad , \\ \delta P_{\lambda}(z,t) &= \delta P_{\lambda}(z)\exp(\lambda t) + \text{c.c.} \quad , \\ \delta\Delta_{\lambda}(z,t) &= \delta\Delta_{\lambda}(z)\exp(\lambda t) + \text{c.c.} \end{aligned} \tag{4.25}$$

Introducing this ansatz (4.25) into (4.23) and taking (4.24) into account, one concludes after some calculations that the possible values for λ are labeled by two indexes n and j (n = 0, ±1, ±2, ...): to be precise, for a given n the values λ_{nj} are the solutions of the eigenvalue equation

$$\lambda = -i\alpha_n + \frac{c}{L}\ln(1 - T) - 2Ck\gamma_{\perp} \quad .$$

$$\frac{1}{L}\int_0^L dz\,\frac{1}{1 + X_{st}^2(z)}\,\frac{\gamma_{\|}[1 - X_{st}^2(z)] + \lambda}{(\gamma_{\perp} + \lambda)(\gamma_{\|} + \lambda) + \gamma_{\perp}\gamma_{\|}X_{st}^2(z)} \tag{4.26}$$

where

$$\alpha_n = 2\pi n\, c/L \quad , \quad n = 0, \pm 1, \pm 2\ \dots \quad . \tag{4.27}$$

The index n labels the modes of the cavity; n = 0 corresponds to the mode which is resonant with the incident field. Since $X_{st}(z)$ is real one easily finds from (4.26) that $\lambda_{(-n)j} = \lambda_{nj}^*$. In the limit $T \ll 1$, $\alpha_{abs}L \ll 1$, with $C = \alpha_{abs}L/2T$ fixed, one can introduce the following simplifications in (4.26): a) $\ln(1 - T) \simeq -T$; and b) the field $X_{st}(z)$ can be taken constant, so that $X_{st}(z) \simeq X_{st}(L) = x$. Hence, as verified by lengthly but elementary calculations, (4.26) reduces to the cubic equation

$$\lambda^3 + (k + \gamma_\perp + \gamma_\parallel + i\alpha_n)\lambda^2 + [k\gamma_\parallel + \gamma_\perp\gamma_\parallel(1 + x^2) + k\gamma_\perp \frac{y}{x} + (\gamma_\perp + \gamma_\parallel)i\alpha_n]\lambda + k\gamma_\perp\gamma_\parallel\left[2x^2 + \frac{y}{x}(1 - x^2)\right] + \gamma_\perp\gamma_\parallel(1 + x^2)i\alpha_n = 0 \quad . \tag{4.28}$$

Therefore in this limit one has three eigenvalues λ_{nj} for each n. For n=0, the roots of (4.28) coincide with the decay constants λ_i discussed in Sect.4.2.1c. This is expected since in the mean field approach only the resonant mode is considered. For n = 0 (resonant mode) we recover the results of the MFT. On the other hand, in the present treatment we consider *all the modes* of the cavity, thereby taking propagation effects into account.

Equation (4.28) also holds for the laser with injected signal [4.62]. In particular, for y = 0 (no external field) (4.28) reduces to the characteristic equation derived in the linear stability analysis of the ring laser [4.63,64]. The stationary state is stable if and only if the roots of (4.28) have a nonpositive real part for all n; in fact, only in this case are all the modes stable. Now, in the limit $T \to 0$ we can neglect all the terms proportional to $k = cT/L$, so that (4.28) reduces to

$$(\lambda + i\alpha_n)[(\lambda + \gamma_\perp)(\lambda + \gamma_\parallel) + \gamma_\perp\gamma_\parallel x^2] = 0 \quad . \tag{4.28'}$$

Therefore for $T \to 0$ the three roots of (4.28) approach the values $\lambda_{n1} = -i\alpha_n$, $\lambda_{n2,3} = (-1/2)\cdot\{\gamma_\perp + \gamma_\parallel \pm[(\gamma_\perp - \gamma_\parallel)^2 - 4\gamma_\perp\gamma_\parallel x^2]^{1/2}\}$. Hence for $T \ll 1$ only λ_{n1} can yield a positive real part. Since $k \propto T$, the second and third terms in the rhs of (4.26) are $\mathcal{O}(T)$, so that λ_{n1} can be approximately calculated by standard perturbative procedure at first order in T, obtaining

$$\lambda_{n1} = - i\alpha_n - k\left[1 + \frac{2C\gamma_\perp}{1 + x^2} \cdot \frac{\gamma_\parallel(1 - x^2) - i\alpha_n}{(\gamma_\perp - i\alpha_n)(\gamma_\parallel - i\alpha_n) + \gamma_\perp\gamma_\parallel x^2}\right] \quad . \tag{4.29}$$

For n = 0, (4.29) reduces to the mean-field result in the good cavity limit (4.18) (in fact, for $T \to 0$ we are automatically in the good cavity case).

Note from (4.29) that $\mathrm{Re}\{\lambda_{n1}\}$ has the structure

$$\mathrm{Re}\{\lambda_{n1}\} = G_n - k \quad ,$$

$$G_n = \frac{-2Ck\gamma_\perp}{1 + x^2}\,\frac{\gamma_\parallel(1 - x^2)[\gamma_\perp\gamma_\parallel(1 + x^2) - \alpha_n^2] + \alpha_n^2(\gamma_\perp + \gamma_\parallel)}{[\gamma_\perp\gamma_\parallel(1 + x^2) - \alpha_n^2]^2 + \alpha_n^2(\gamma_\perp + \gamma_\parallel)^2} \quad . \tag{4.29'}$$

When G_n is positive this is a gain-minus-loss form. Hence, when $\mathrm{Re}\{\lambda_{n1}\} > 0$ the gain exceeds the loss so that our passive system behaves like a laser with respect to the unstable modes. However, different from the usual laser threshold which shows

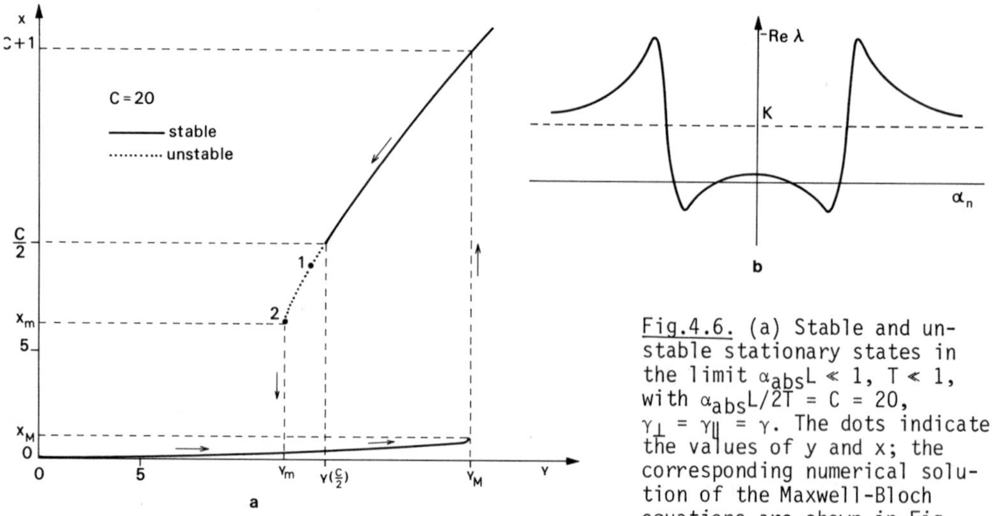

Fig.4.6. (a) Stable and unstable stationary states in the limit $\alpha_{abs}L \ll 1$, $T \ll 1$, with $\alpha_{abs}L/2T = C = 20$, $\gamma_\perp = \gamma_\parallel = \gamma$. The dots indicate the values of y and x; the corresponding numerical solution of the Maxwell-Bloch equations are shown in Fig. 4.8. (b) Qualitative shape of the graph of $-\mathrm{Re}\{\lambda_{n1}\}$ (4.29) vs $\alpha_n = 2\pi cn/L$ in correspondence to the dotted (unstable) part of the hysteresis cycle in (a)

a second-order phase transition, this instability can also lead to a first-order phase transition.

A typical plot of $-\mathrm{Re}\{\lambda_{n1}\}$ is shown in Fig.4.6b. The analysis of the instability condition $\mathrm{Re}\{\lambda_{n1}\} > 0$ leads to a biquadratic equation, the discussion of which yields the following conclusion. The stationary state is unstable when the following two conditions are simultaneously satisfied:

$$R \geq 0 \quad , \qquad S + R^{1/2} \geq 0 \quad ,$$

$$\begin{aligned} R = {} & \gamma_\perp^2\gamma_\parallel^2 x^4\left(1 - \frac{y}{x}\right)^2 + \left(\gamma_\parallel^2 - \gamma_\perp^2 \frac{y}{x}\right)^2 \\ & -2\gamma_\perp\gamma_\parallel x^2\left[3\gamma_\parallel^2 + 4\gamma_\perp\gamma_\parallel + \frac{y}{x}\left(3\gamma_\perp^2 - \gamma_\parallel^2\right) + \gamma_\perp^2\frac{y^2}{x^2}\right] \quad , \end{aligned} \tag{4.30}$$

$$S = \gamma_\parallel\left(3\gamma_\perp x^2 - \gamma_\parallel\right) - \frac{y}{x}\,\gamma_\perp\left(\gamma_\perp + \gamma_\parallel x^2\right) \quad ,$$

provided that at least one of the discrete values α_n lies in the interval $\alpha_{min} < \alpha_n < \alpha_{max}$ where

$$\alpha_{\substack{max\\min}} = \frac{1}{\sqrt{2}}\left(S \pm \sqrt{R}\right)^{1/2} \quad . \tag{4.31}$$

The points of the curve y(x) on the part with negative slope are always unstable because the resonant mode is unstable (as one sees immediately from the expression $\lambda_{01} = -k\,dy/dx$). On the other hand, under suitable conditions it happens that in correspondence to a part of the plot y(x) *with positive slope* some off-resonance modes are unstable. For the sake of definiteness, let us analyze the case $\gamma_\perp = \gamma_\parallel \overset{\mathrm{def}}{=} \gamma$. The resulting picture is the following:

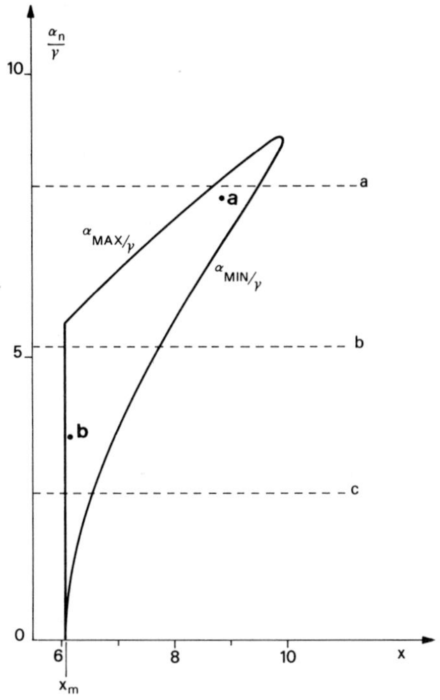

Fig.4.7. Plot of the functions α_{max} and α_{min} which shows the frequency interval for instability for $\gamma_\perp=\gamma_\parallel=\gamma$. The points indicate the values of the parameter $\alpha_1/\gamma=2\pi c/L\gamma$ in correspondence to the two graphs in Fig.4.8

1) For $C < 2(1+\sqrt{2})$ all the points of the plot $x = x(y)$ which lie on the part with positive slope are stable.

2) For $C > 2(1+\sqrt{2})$ the points in the one-atom branch (high-transmission branch) such that $x < C/2$ (Fig.4.6a) are unstable provided at least one of the discrete values α_n lies in the range $\alpha_{min} < \alpha_n < \alpha_{max}$ where (Fig.4.7)

$$\alpha_{\substack{max\\min}} = \gamma\left(x^2 - C - 1 \pm \sqrt{C^2 - 4x^2}\right)^{1/2} \quad . \tag{4.32}$$

e) *Self-Pulsing in Optical Bistability*

The analysis of the previous section strongly suggests the possibility of self-pulsing in OB, similar to what is found in ring lasers [4.63,64]. In fact, in correspondence to an unstable point one has two possibilities: for a fixed y either the system jumps to the corresponding steady state in the cooperative branch (low-transmission branch), which is always stable, or the system evolves to a time periodic state (limit cycle), so that the transmitted light becomes a sequence of short pulses (self-pulsing). The numerical solution of the MBE (4.1) shows that both these possibilities occur. Hence, the analysis of [4.18] showed for the first time the rise of self-pulsing in a purely passive system.

In principle there is a third possibility, namely that the system evolves to a chaotic situation in which it exhibits a completely irregular sequence of pulses.

An example of this behavior is given in [4.20] for a case of dispersive multistability. The nature of this behavior is completely different from that of our self-pulsing behavior. In fact, the instability that leads to chaos in [4.20] does not arise in the absorptive case.

In our calculations we have fixed $\gamma_\perp = \gamma_\| = \gamma$, $C = 20$, $T = 0.1$, and $\mathcal{L} = 5L$. Time is expressed in units L/c. In all the cases that we have considered, the modes $n = \pm 1$ (i.e., the modes of frequency $\omega_0 \pm 2\pi c/\mathcal{L}$) are unstable according to (4.29). We have taken the initial condition for the MBE in such a way that the deviations δE, δP, $\delta\Delta$ from the unstable stationary solution are initially small and only the modes $n = \pm 1$ are initially excited. Typical results are shown in Fig.4.8, where one sees the envelope of the time evolution of the transmitted field. The initial stage of the evolution, in which the deviation from the unstable steady state is exponentially amplified, agrees with the predictions of the linearized MBE (4.23). In the case of Fig.4.8a the system evolves towards a limit cycle. The frequency of the oscillations in such a cycle is roughly equal to the off-resonance $2\pi c/\mathcal{L}$ of the unstable modes $n = \pm 1$ within an error proportional to T, i.e., of the order of 10%. The mean value of the oscillations is always lower than the unstable steady-state value, showing a kind of attractive force exerted by the stable steady state. Note that when the system exhibits continuous pulsations it works like a parametric oscillator which coherently transfers energy from the external signal to some unstable mode whose frequency α_n is alway smaller than the Rabi frequency γx (4.32). This instability can be roughly understood as the combined effect of the small signal gain for a coherently saturated absorber without cavity [4.2,65] (Fig.4.6b) and the cavity boundary conditions which provide feedback and loss mechanisms. These boundary conditions provide half the physics of the problem since the field acting on the atoms is not the external driving field as in [4.2,65] but a discontinuous function of it. In fact, the instability condition $C > 2(1 + \sqrt{2})$ implies the validity of the bistability condition $C > 4$.

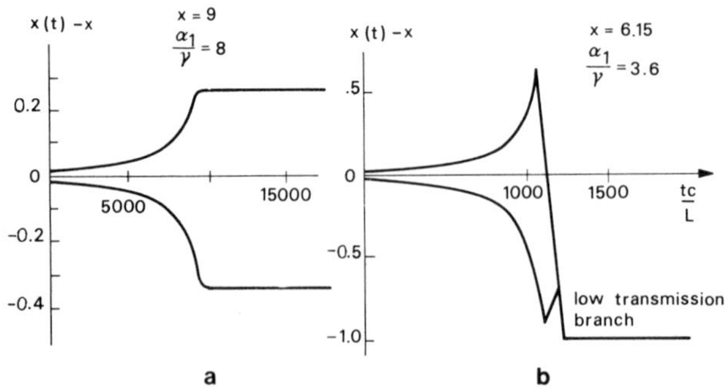

Fig.4.8a,b. Envelope of the time evolution of the transmitted field for $C = 20$, $\gamma_\perp = \gamma_\| = \gamma$, $\mathcal{L} = 5L$, $T = 0.1$. The points of the xy plane corresponding to (a,b) are indicated in Fig.~~4.6a~~ 4.7

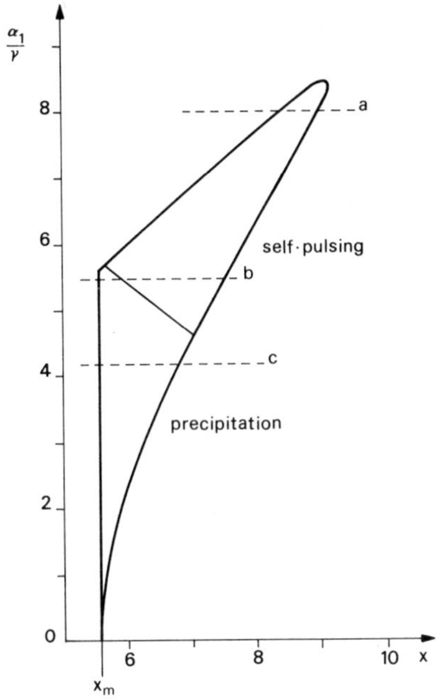

Fig.4.9. Subdivision of the instability region into self-pulsing and precipitation regimes. C = 20, T = 0.1, L = 5L, $\gamma_\perp = \gamma_\parallel = \gamma$

In the case of Fig.4.8b the system precipitates to the low-transmission steady state. Clearly in this situation the attraction of the lower steady state becomes overwhelming. The subdivision of the instability region of Fig.4.7 into self-pulsing and precipitation regimes is shown in Fig.4.9 [4.66].

Let us now describe what happens when the incident field y along the high intensity branch decreases, starting from a value of y such that $x > C/2$ (Fig.4.9) [4.66]. As y decreases, x decreases whereas $\alpha_1 = 2\pi c/L$ remains constant. Hence the point $x, \alpha_1/\gamma$ moves in the plane of Fig.4.9 along a horizontal line as a, b, c from the right to the left. Let us first consider the case of the line a. When the point $\{x, \alpha_1/\gamma\}$ enters from the right into the instability region bounded by the lines α_{max}/γ, α_{min}/γ, $x = x_m$ the self-pulsing behavior appears *abruptly*, with oscillations of finite amplitude. In other words, crossing the right boundary of the instability region the system shows a first-order-like phase transition from stationary to self-pulsing behavior. When y (or x) is decreased, the amplitude of the oscillations continuously decreases until, in correspondence to the left boundary of the instability region, the oscillations vanish and the system is again back at a stationary state. Hence crossing the left boundary shows a second-order-like phase transition from self-pulsing to stationary behavior. Let us now consider somewhat larger values of L, as in the case of the line b. When we cross the left boundary everything happens as in the case of the line a. However, before arriving at the left boundary, the system precipitates to the low-transmission branch. Finally, in the case of line c the system precipates as soon as one enters into the instability region.

An analytical theory of self-pulsing in absorptive OB has been elaborated in [4.67-69]. This theory is based on HAKEN's theory of generalized Ginzburg-Landau equations for phase-transition-like phenomena in systems far from thermal equilibrium [4.70-72]. In [4.67] HAKEN's formalism is simplified and generalized to the case of stationary state nonuniform in space, as in OB. In [4.68] the stable modes are adiabatically eliminated following the iterative procedure of [4.71]. This procedure turns out to reproduce satisfactorily the numerical results from the Maxwell-Bloch equations, but only in the case of a second-order phase transition. Finally in [4.69] the adiabatic elimination is performed exactly in the mean-field limit, thereby obtaining a description which also works when the amplitude of the oscillations is quite large, as in the case of a first-order phase transition. The results of [4.69] lead to the prediction of new types of histeresis cycles which involve both CW and self-pulsing states.

We briefly mention that the self-pulsing instabilities in the mixed absorptive-dispersive case (ring cavity) have been analyzed in [4.73]. In [4.74,75] the same problem is considered for absorptive OB in a Fabry-Perot.

4.2.2 Quantum-Statistical Theory

To describe the spectrum of transmitted and fluorescent light or the photon statistics of the transmitted field one must deal with the fluctuations of the system. To this aim let us consider the quantum-statistical formulation of the mean-field model [4.30], which is a straightforward generalization of the well-known one-mode laser model [4.76,77].

Let r_i^+ and r_i^- be the raising and lowering operators of the ith two-level atom in the cavity ($i = 1, \ldots N$) and let $r_{3i} = (1/2)(r_i^+ r_i^- - r_i^- r_i^+)$ be the population inversion operator of the ith atom. The collection of N atoms is described by the total population inversion operator

$$R_3 = \sum_{i=1}^{N} r_{3i} \tag{4.33}$$

and by the collective dipole operators $R^\pm$ defined as

$$R^\pm = \sum_{i=1}^{N} r_i^\pm \exp(\pm i k_0 \cdot x_i) \quad , \tag{4.34}$$

where k_0 is the wave vector of the injected field and x_i is the position of the ith atom. The operators $R^\pm$ and R_3 obey the angular momentum commutation relations

$$[R^+,R^-] = 2R_3 \quad , \quad [R_3,R^\pm] = \pm R^\pm \quad . \tag{4.35}$$

Let A be the annihilation operator of photons in the mode of the cavity which is resonant with the incident field. Let us consider the statistical operator W(t) of the coupled system atoms plus resonant radiation mode. In the interaction representation, W(t) obeys the master equation (ME)

$$\frac{dW}{dt} = - iL_{AF}W + \Lambda_F W + \Lambda_A W \quad , \tag{4.36}$$

where

$$L_{AF}W = \hbar[H_{AF},W] \quad , \quad H_{AF} = i\bar{g}(A^+R^- - AR^+) \quad , \tag{4.37}$$

$$\Lambda_F W = k\{[(A - \alpha), W(A - \alpha)^+] + [(A - \alpha)W , (A - \alpha)^+]\} \quad , \tag{4.38}$$

$$\Lambda_A W = \sum_{i=1}^{N} \left\{\frac{\gamma_\parallel}{2} [(r_i^-,Wr_i^+) + (r_i^-W,r_i^+)] \right.$$

$$\left. + \left(\gamma_\perp - \frac{\gamma_\parallel}{2}\right)[(r_{3i},Wr_{3i}) + (r_{3i}W,r_{3i})]\right\} \quad . \tag{4.39}$$

In (4.37) $\bar{g}$ is a suitable coupling constant in the dipole approximation

$$\bar{g} = \left(\frac{2\mu\hbar\omega_0}{V}\frac{L}{\mathcal{L}}\right)^{1/2} \mu \quad ; \tag{4.40}$$

in (4.38) α, which is assumed real and positive, is proportional to the injected field amplitude

$$\alpha = (V\mathcal{L}/8\pi\hbar\omega_0 L)^{1/2}(E_I/\sqrt{T}) \quad . \tag{4.41}$$

Accordingly, the mean value $\langle A\rangle$ is proportional to the internal field amplitude E of (4.14):

$$\langle A\rangle = (V\mathcal{L}/8\pi\hbar\omega_0 L)^{1/2}E = \sqrt{N_S}x \quad , \tag{4.42}$$

where N_S is the saturation photon number

$$N_S = \gamma_\perp\gamma_\parallel/4\bar{g}^2 \quad . \tag{4.42'}$$

The term L_{AF} describes the interaction between the atoms and the field in the dipole and rotating-wave approximations. The part Λ_F describes the escape of photons from the active volume and takes into account the presence of the incident field. The structure of (4.38) can be understood as follows: the steady state for the field statistical operator W_F in the absence of atoms is given by the solution of the equation $\Lambda_F W_F = 0$. Such a solution is the coherent state $W_F = |\alpha\rangle\langle\alpha|$, which is obviously the statistical operator of the classical incident field. Finally, the part Λ_A arises from the radiative and collisional decay of the atoms. The part of Λ_A proportional to $\gamma_\perp - (\gamma_\parallel/2)$ is a dephasing term. In comparison with the laser, Λ_A does not describe any upward transition because in the case of optical bistability there is no pump action.

Let us now show that in the semiclassical approximation the master equation (4.36) reproduces the semiclassical mean-field model (4.14). We consider the equations of motion for the mean values of R^-, R_3 and A which follow from (4.36) using the definitions $\langle R^-\rangle(t) = \mathrm{trace}\,[R^-W(t)]$, etc. We obtain

$$\langle \dot{R}^-\rangle = 2\bar{g}\langle AR_3\rangle - \gamma_\perp \langle R^-\rangle \ , \tag{4.43a}$$

$$\langle \dot{R}_3\rangle = -\bar{g}(\langle AR^+\rangle + \langle A^+R^-\rangle) - \gamma_\parallel(\langle R_3\rangle + \frac{N}{2}) \ , \tag{4.43b}$$

$$\langle \dot{A}\rangle = \bar{g}\langle R^-\rangle - k(\langle A\rangle - \alpha) \ . \tag{4.43c}$$

Now we factorize the mean values of products into the products of mean values. Using (4.2) and (4.40-42), and introducing the definitions

$$P = -\langle R^-\rangle \quad , \quad \Delta = -\langle R_3\rangle \quad , \tag{4.44}$$

one easily verifies that (4.43) are equivalent to (4.14).

A many-mode master equation for OB, which holds in the mean-field limit both in the absorptive and in the dispersive case, has been formulated in [4.78].

a) *Spectrum of Transmitted Light*

The spectrum $S(\omega)$ of the transmitted light is given by the Fourier transform of the time correlation function at steady state $\langle A^+(t)A\rangle_{st}$:

$$S(\omega) = \frac{1}{\pi}\,\mathrm{Re}\left\{\int_0^\infty dt\, \exp[-i(\omega - \omega_0)t]\langle A^+(t)A\rangle_{st}\right\} \ . \tag{4.45}$$

Hence to obtain the spectrum one must calculate the fluctuations of the system around the steady state. More specifically, for any given incident field y let us choose one of the two stable steady states and let us call x the transmitted field in the chosen state. Subdividing A(t) into the stationary mean value $\langle A\rangle_{st} = \sqrt{N_s}x$ cf. (4.42,42') and the fluctuation $\delta A(t) = A(t) - A_{st}$, we have that $S(\omega)$ is composed by a coherent and an incoherent part,

$$S(\omega) = S_{coh}(\omega) + S_{inc}(\omega) \ , \tag{4.46}$$

$$S_{coh}(\omega) = N_S x^2 \delta(\omega - \omega_0) \ , \tag{4.47}$$

$$S_{inc}(\omega) = \frac{1}{\pi}\,\mathrm{Re}\left\{\int_0^\infty dt\, \exp[-i(\omega - \omega_0)t]\ \langle \delta A^+(t)\delta A\rangle_{st}\right\} \ . \tag{4.48}$$

The coherent or classical part has the same frequency of the injected field and is proportional to the intensity x^2 of the transmitted field. The incoherent part is the quantum-mechanical contribution and arises from the fluctuations around the steady state. In order to calculate $S_{inc}(\omega)$, one translates the master equation (4.36) into a classical-looking Fokker-Planck equation in five macroscopic variables which correspond to polarization, population inversion, electric field [4.79]. By linearizing this equation and using the so-called *regression theorem* one finds [4.44] that $S_{inc}(\omega)$ is given by the superposition of few Lorentzians which are peaked at $\omega = \omega_0 + \mathrm{Im}\lambda_i$ and have width $\mathrm{Re}\{\lambda_i\}$, where λ_i are the eigenvalues of the linearized semiclassical equations (4.14) (cf. Sect.4.2.1c). Hence one understands why most of the features of the spectrum can be predicted by simply analyzing the semiclassical equations, as was done in [4.30]. In [4.44] the spectrum is calculated in the bad cavity limit $k \gg \gamma_\perp, \gamma_\parallel$ and in the opposite good cavity situation $k \ll \gamma_\perp, \gamma_\parallel$. Let us consider the two cases separately.

1) $k \gg \gamma_\perp, \gamma_\parallel$ (bad cavity case). In this case, the relevant eigenvalues are

$$\lambda_\pm = \frac{\gamma_\perp}{2}\left\{\frac{\gamma_\parallel}{\gamma_\perp} + \frac{y}{x} \pm \left[\left(\frac{\gamma_\parallel}{\gamma_\perp} - \frac{y}{x}\right)^2 - 4\frac{\gamma_\parallel}{\gamma_\perp}x(2x - y)\right]^{1/2}\right\} , \tag{4.49}$$

$$\lambda_3 = \gamma_\perp \frac{y}{x} . \tag{4.50}$$

The root λ_- coincides with the damping constant $\bar{\lambda}$ discussed in Sect.4.2.1c; when x,y lies on the cooperative branch λ_- is well approximated by expression (4.19). When $\lambda_\pm$ are real $S_{inc}(\omega)$ is given by the superposition of three Lorentzians,

$$S_{inc}(\omega) = \frac{C\gamma_\perp^2}{\pi k}\left[w_3\frac{\lambda_3}{(\omega - \omega_0)^2 + \lambda_3^2} + w_-\frac{\lambda_-}{(\omega - \omega_0)^2 + \lambda_-^2} + w_+\frac{\lambda_+}{(\omega - \omega_0)^2 + \lambda_+^2}\right] , \tag{4.51}$$

where

$$w_3 = \frac{1}{2\lambda_3}\frac{x^2}{1 + x^2} ,$$

$$w_\pm = \mp\frac{1}{2\lambda_\pm}\frac{x^2}{1 + x^2}\frac{1}{\lambda_+^2 - \lambda_-^2}\cdot\left[\left(\frac{\gamma_\parallel}{\gamma_\perp} - 1\right)\lambda_\pm^2 - \gamma_\parallel\left(\frac{\gamma_\parallel}{\gamma_\perp} - 1 - 2x^2\right)\right] . \tag{4.52}$$

When $\lambda_\pm$ are complex conjugate one puts

$$\lambda_\pm = \lambda_1 \pm i\lambda_2 \tag{4.53}$$

and obtains

$$S_{inc}(\omega) = \frac{C\gamma_\perp^2}{\pi k}\left[w_3\frac{\lambda_3}{(\omega - \omega_0)^2 + \lambda_3^2} + g(\omega - \omega_0) + g(\omega_0 - \omega)\right] , \tag{4.54}$$

where

$$g(\nu) = \frac{1}{8}\frac{1}{\lambda_1^2 + \lambda_2^2}\frac{\gamma_\perp^3}{(\nu - \lambda_2)^2 + \lambda_1^2}\left\{2\left(\frac{\gamma_\parallel}{\gamma_\perp}\right)^2\left(2x^2 + 1 - \frac{\gamma_\parallel}{\gamma_\perp}\right)\right.$$

$$-\frac{\nu}{\gamma_\perp\lambda_2}\left[\left(\frac{\gamma_\parallel}{\gamma_\perp}-1\right)(\lambda_1^2+\lambda_2^2)-\gamma_\parallel^2\left(\frac{\gamma_\parallel}{\gamma_\perp}-1-2x^2\right)\right]\Bigg\}\frac{x^2}{1+x^2} \tag{4.55}$$

For $\gamma_\parallel = 2\gamma_\perp$ (4.51) and (4.54) coincide with the formulas independently derived by AGARWAL et al. via quantum-mechanical Langevin equations [4.42,43]. The hysteresis cycle of the spectrum for $2\gamma_\perp = \gamma_\parallel = \gamma$ is shown in Fig.4.10 for $C \gg 1$. When the system is on the cooperative branch, $\lambda_\pm$ are real, so that the spectrum $S_{inc}(\omega)$ is a single line. For $y \ll C$ one has

$$\lambda_3 \simeq \lambda_+ \simeq C\gamma \quad , \quad \lambda_- \simeq \gamma \quad . \tag{4.56}$$

In these conditions the contribution of the term proportional to w_- in (4.51) is negligible, so that the spectrum is a broad line whose halfwidth is γC, which coincides with the cooperative linewidth γ_R of pure superfluorescence [4.60]. Since $C \propto N$, in this situation the linewidth is *proportional to the number of atoms* (cooperative line broadening). Increasing y along the cooperative branch, the peak corresponding to the soft mode $\bar{\lambda}$ emerges from the cooperative background (Fig.4.10b). Approaching the upper bistability threshold the spectrum becomes a narrow line in which the soft mode dominates and the cooperative background is completely negligible (Fig.4.10c,d). This *line narrowing* (halfwidth $\lambda_- \to 0$) is clearly a manifestation of the critical slowing down illustrated in Sect.4.2.1c.

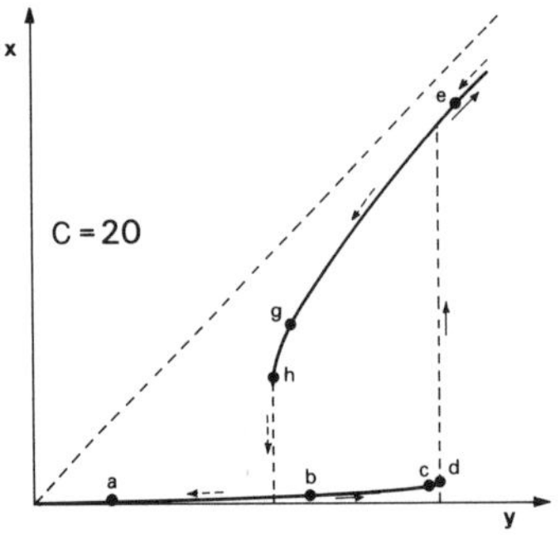

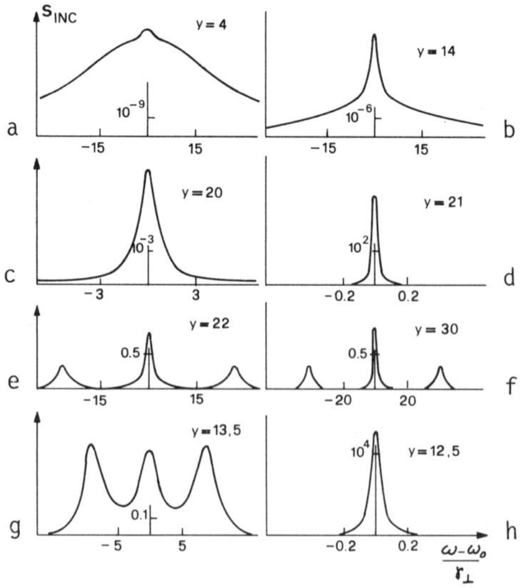

Fig.4.10a-h. Hysteresis cycle of the incoherent part of the spectrum of transmitted light for C = 20 and $\gamma_\parallel = 2\gamma_\perp \ll k$. S_{inc} is given in units $C/2\pi k$. The scale varies from diagram to diagram as indicated

Let us now cross the threshold $y = y_M$ so that the system jumps to the one-atom branch. The roots $\lambda_\pm$ are complex conjugate, so that the spectrum suddenly becomes a triplet (Fig.4.10e). This means a discontinuous appearance of a dynamical Stark effect. For $y \gg y_M$ one has $x \simeq y$ and

$$\lambda_3 \simeq \frac{\gamma}{2} \quad , \quad \lambda_1 \simeq \frac{3}{4}\gamma \quad , \quad \lambda_2 \simeq \gamma \frac{y}{\sqrt{2}} \quad . \tag{4.57}$$

Hence since $\lambda_2 \gg \lambda_1$ and $x \gg 1$, $S_{inc}(\omega)$ takes the simple form

$$S_{inc}(\omega) \propto \frac{\gamma/2}{(\omega - \omega_0)^2 + (\gamma/4)^2} + \frac{1}{2}\left[\frac{\frac{3}{4}\gamma}{(\omega - \omega_0 - \Omega_I)^2 + \frac{9}{16}\gamma^2} + \frac{\frac{3}{4}\gamma}{(\omega - \omega_0 + \Omega_I)^2 + \frac{9}{16}\gamma^2}\right] \quad , \tag{4.58}$$

where Ω_I is the Rabi frequency of the incident field cf. (4.21) . Equation (4.52) coincides with the lineshape predicted for the spectrum of fluorescent light in the high-intensity situation by the one-atom theory of resonance fluorescence [4.80,81]. For $C \gg 1$, $\Omega_I \gg \gamma$ so that the sidebands are well separated from the central line.

Let us now decrease y along the one-atom branch. The two sidebands get nearer and nearer to the central line (Fig.4.10g) until in the vicinity of the lower threshold $y = y_m$ the root $\lambda_\pm$ becomes real. There is again line narrowing because the linewidth $2\lambda_-$ tends to zero.

2) $k \ll \gamma_\perp, \gamma_\parallel$ (good cavity case). In this case, the relevant eigenvalues are $\bar{\lambda}$ given by (4.18) and

$$\lambda_\varphi = k\,\frac{y}{x} = k\left[1 + 2C\,\frac{1}{1 + x^2}\right] \quad . \tag{4.59}$$

One finds [4.44]

$$S_{inc}(\omega) = \frac{Ck}{2\pi}\,\frac{x^2}{1 + x^2}\left[\frac{2x^2 + 1 - \frac{\gamma_\parallel}{\gamma_\perp}}{(1 + x^2)^2} \cdot \frac{1}{(\omega - \omega_0)^2 + \bar{\lambda}^2} + \frac{1}{(\omega - \omega_0)^2 + \lambda_\varphi^2}\right] \tag{4.60}$$

The eigenvalues $\bar{\lambda}$ and λ_φ are always real, so that for $k \ll \gamma_\perp, \gamma_\parallel$ *one never finds a dynamical Stark effect.* For $y \ll C$ one has

$$\lambda_\varphi \simeq \bar{\lambda} \simeq 2Ck \quad . \tag{4.61}$$

Hence the width of the spectrum is 4Ck, much larger than the empty cavity with 2k (cooperative broadening effect). As in the bad cavity case, for small incident field the linewidth is proportional to N. Approaching the upper bistability threshold $y = y_M$ we find the usual cooperative line narrowing. Crossing the threshold, the spectrum changes discontinuously from a narrow line to a line whose width coincides with the empty cavity width 2k because $\lambda_\varphi \simeq \bar{\lambda} \simeq k$ for $y \gtrsim y_M$. Finally decreasing y along the one-atom branch as usually there is line narrowing at $y = y_m$.

The spectrum of transmitted light, including all the longitudinal modes of the cavity, is given in [4.69]. In this paper the behavior of this spectrum is also described when we approach the self-pulsing instability.

The spectrum of fluorescent light diffused at 90° is analyzed in [4.82,83].

b) *Photon Statistics of the Transmitted Light*

As we anticipated, OB is an example of first-order-like phase transition in an open system far from thermal equilibrium. As is well known, this behavior can also be shown by other systems containing a saturable absorber: a parametrically excited subharmonic oscillator [4.84], laser with saturable absorber [4.85-88] dye laser [4.89,90], and bidirectional ring cavity [4.91]. The characteristic feature of OB with respect to these systems is that it occurs in a purely passive system and that OB never exhibits a second-order transition. Hence OB plays the role of a prototype of first-order transitions in optical systems, exactly as the usual laser with active atoms is just the prototype of second-order phase transitions [4.92,93]. To work out this analogy, one must analyze in full detail the fluctuations of the system. In fact, in the bistable situation only one of the two stationary solutions is absolutely stable, while the other is only metastable. The semiclassical treatment is unable to tell us which one of the two is absolutely stable. In fact, the linear stability analysis checks the stability of the stationary solutions only against the "small" fluctuations around each steady state. Also in the previous section in which we studied the spectrum of the transmitted light, we have only analyzed the small fluctuations around the stationary solutions, thereby treating stable and metastable states on the same footing. However, the system can also develop "large" fluctuations which make the system "tunnel" from the metastable to the stable solution. The probability of such large fluctuations is extremely small, as one sees by applying the method of KRAMERS [4.94,95]; however to analyze the thermodynamic stability of the steady states one must develop a treatment which works out the full spectrum of fluctuations. This treatment is given in [4.36] for the good quality cavity case $k \ll \gamma_\perp, \gamma_\|$. Following the theory of open systems of [4.96] we derive from the ME (4.36) the following Fokker-Planck equation for the Glauber distribution P_G of the transmitted field:

$$\frac{\partial}{\partial t} P_G(x,\varphi,t) = k\left[\frac{\partial}{\partial x}\left(x - y\cos\varphi + \frac{2Cx}{1+x^2} + q\frac{\partial}{\partial x}\frac{x^2}{(1+x^2)^2}\right) + \frac{\partial}{\partial\varphi}\left(\frac{y}{x}\sin\varphi + q\frac{\partial}{\partial\varphi}\frac{1}{1+x^2}\right)\right]P_G(x,\varphi,t) \quad . \tag{4.62}$$

Equation (4.62) is obtained as the Fokker-Planck approximation of an equation which contains derivatives of all orders in x. Here x is a stochastic variable which corresponds to the normalized amplitude of the field, and φ is the phase of the field. The mean value $\langle A\rangle$ at time t is given by

$$\langle A\rangle(t) = \sqrt{N_S}\langle x\, e^{i\varphi}\rangle(t) = \int_0^\infty dx \int_0^{2\pi} d\varphi P_G(x,\varphi,t)x\, e^{i\varphi} \quad . \tag{4.63}$$

The diffusion constant q is given by

$$q = C/2N_S \quad . \tag{4.64}$$

The main feature of (4.62) with respect to similar Fokker-Planck equations for the usual laser [4.97] is that the diffusion terms are intensity dependent. This shows that in OB saturation effects are important not only in the drift motion, but also in the fluctuations.

Let us discuss (4.62) at steady state ($\partial P_G/\partial t = 0$). It does not appear easy to obtain the exact stationary solution because it depends on both amplitude and phase. However, a very well approximated expression for the amplitude stationary distribution can be easily obtained in the following way. At a semiclassical level, the phase has *only one* stationary value, $\varphi = 0$. At a quantum-statistical level, the phase will fluctuate around $\varphi = 0$ but these fluctuations are small because the diffusion constant q is quite small. Hence at steady state one can linearize (4.62) with *respect to the phase only*, so that $\cos\varphi$ is simply replaced by 1. At this point, one can integrate (4.62) with respect to the phase obtaining the following closed equation for the amplitude distribution $P(x) = \int d\varphi P_G(x,\varphi)$:

$$\left(x - y + \frac{2Cx}{1 + x^2} + q\frac{\partial}{\partial x}\,\frac{x^2}{(1 + x^2)^2}\right)P(x) = 0 \quad . \tag{4.65}$$

The solution of (4.65) is

$$P(x) = N[(1 + x^2)/x]^2 \exp\left[-\frac{1}{q}V(x)\right] \quad , \tag{4.66}$$

where N is a suitable normalization constant and

$$V(x) = \int dx\left(\frac{1 + x^2}{x}\right)^2\left(x - y + \frac{2Cx}{1 + x^2}\right)$$
$$= (2C + 1)\ln x + (x - y)^2 + \frac{y}{x} + x^2\left(C - \frac{1}{3}yx + \frac{1}{4}x^2\right) \quad . \tag{4.67}$$

The potential V(x) plays the role of a generalized free energy in our problem. Clearly equation $dV/dx = 0$, which determines the extrema of the potential, coincides with the semiclassical state equation (4.12). For $q \ll 1$, the factor $[(1 + x^2)/x]^2$ produces a negligible shift in the position of the extrema of distribution P(x) which then coincide with the extrema of V(x). Hence the stable semiclassical solutions correspond to most probable values (i.e., peaks of the distribution function), while the unstable solutions correspond to least probable values. In particular for $C > 4$ in the bistable situation $y_m < y < y_M$ (Fig.4.5) P(x) has two peaks at $x = x_1$ and $x = x_3$. The parameter q controls the width of the peaks: the smaller is q, the narrower are the peaks. The smallness of q also has another important consequence. The range of values of y in which the two peaks have comparable areas is very small, i.e., in the largest part of the bistable region

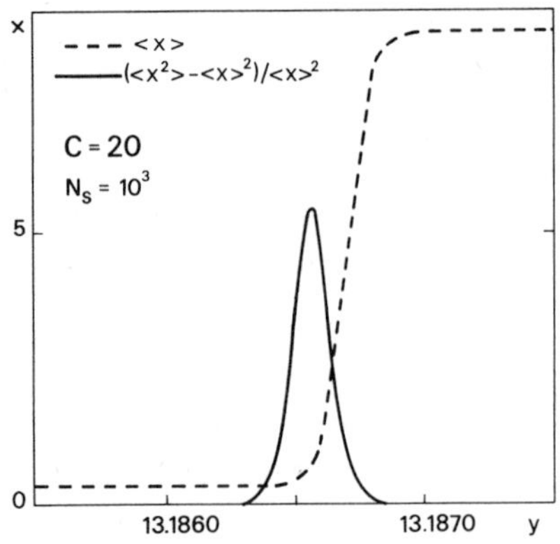

Fig.4.11. Mean value and relative fluctuation of the transmitted field as calculated from [4.66]

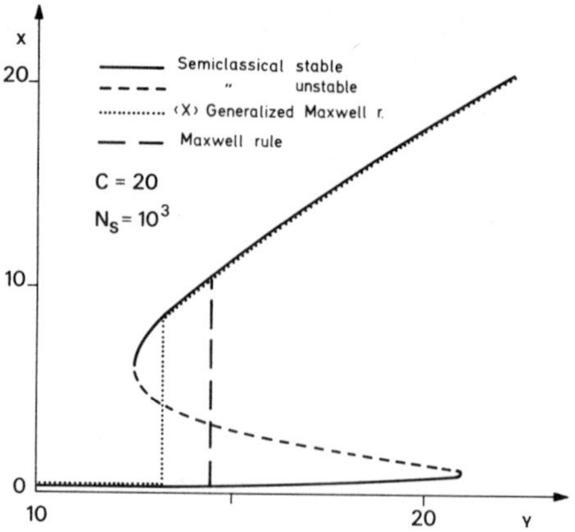

Fig.4.12. Semiclassical stationary solutions, Maxwell rule, and mean value of the normalized field amplitude x as calculated from [4.66]

$y_m < y < y_M$ one of the two peaks dominates the other. Figure 4.11 shows the mean value $\langle x \rangle$ of the amplitude and the relative fluctuation $(\langle x^2 \rangle - \langle x \rangle^2)/\langle x \rangle^2$ for $C = 20$ and $q = 10^{-2}$. On the other hand, Fig.4.12 compares the mean value $\langle x \rangle$ with the semiclassical solutions. Clearly the mean value coincides with one of the two stable semiclassical solutions everywhere except in a narrow transition region in which the two peaks have comparable areas.

By means of Fig.4.12 we can now decide which one of the two semiclassical solutions, which are stable according to the linear stability analysis, is absolutely stable and which one is metastable. In fact, the absolutely stable solution is the one which practically coincides with $\langle x \rangle$. As we see from Fig.4.11 the fluctuations are always very small except in the transition region where we find a remarkable peak which arises from the strong competition between two peaks of comparable areas.

Clearly the mean value exhibits a behavior which strongly resembles a first-order phase transition. The smaller q is, the sharper the transition. In fact, let us consider the *thermodynamic limit* $N \to \infty$, $V \to \infty$ with $\rho = N/V$ constant. In this limit q tends to zero and one finds the discontinuous transition [4.36]

$$\langle x \rangle \xrightarrow[q \to 0]{} \begin{cases} x_1 & \text{for} \quad V(x_1) < V(x_3) \\ x_3 & \text{for} \quad V(x_1) > V(x_3) \quad . \end{cases} \tag{4.68}$$

Equation (4.68) justifies the choice of V(x) as free energy, and provides a *generalized Maxwell rule* for our problem. Figure 4.12 shows that this rule is quite different from the usual Maxwell rule of equilibrium thermodynamics, which prescribes cutting the semiclassical curve in the x,y plane in such a way that one obtains two regions of equal areas. A similar phenomenon arises in chemical reactions [4.98]. It is easy to verify that this discrepancy is due to the fact that the diffusion coefficients in (4.62) are not constant. In fact, it has been shown [4.36] that for constant diffusion one has the usual Maxwell rule.

The amplitude fluctuations in the incident field have been studied in [4.38]. The approach to the stationary solution (4.66), including the tunnelling process, has been analyzed in [4.95,99].

4.3 Theory of Mixed Absorptive-Dispersive OB in a Ring Cavity

The exact theory of OB in a ring cavity has been generalized to take into account the effect of cavity mistuning, atomic detuning, and inhomogeneous broadening [4.19]. The main result of this treatment is the relation which at steady state links the normalized incident and transmitted intensities, defined as

$$Y = \left(\frac{\mu E_I}{h}\right)^2 \Big/ \gamma_\perp \gamma_\parallel \, T = y^2 \quad ,$$

$$I_T = \left(\frac{\mu |E_T|}{\hbar}\right)^2 / \gamma_\perp \gamma_\parallel \, T \quad . \tag{4.69}$$

In the case of homogeneous broadening the function $I_T(Y)$ is expressed in parametric form as follows:

$$I_T = \frac{2}{\rho^2 - 1} [\alpha_{abs} L - (1 + \Delta^2) \ln\rho] \quad , \tag{4.70a}$$

$$Y = \frac{2}{T^2} \frac{\alpha_{abs} L - (1 + \Delta^2) \ln\rho}{\rho^2 - 1} \quad . \tag{4.70b}$$

$$[\rho^2 + R^2 - 2R\rho\cos(\Delta \ln\rho - \theta T)] \quad ,$$

where the parameter $\rho = |X(0)/X(L)|$[cf. (4.4c)]. The atomic detuning is Δ

$$\Delta = (\omega_A - \omega_0)/\gamma_\perp \quad , \tag{4.71}$$

where ω_A is the central frequency of the atomic line and ω_0 the frequency of the incident field. The cavity mistuning parameter θ is,

$$\theta = (\omega_c - \omega_0)/k \quad , \tag{4.72}$$

where ω_c is the frequency of the cavity that is nearest to ω_0. In the purely absorptive case $\theta = \Delta = 0$, $I_T = x^2$, so that (4.70a) and (4.70b) reduce to (4.9) and (4.8), respectively. In the mean-field limit (4.11) with θ constant, (4.70) reduce to [4.47,50]

$$Y = I_A\left[\left(1 + \frac{2C}{\Delta^2 + 1 + I_T}\right)^2 + \left(\theta - \frac{2C\Delta}{\Delta^2 + 1 + I_T}\right)^2\right] \quad . \tag{4.73}$$

The analytical relation (4.70b) has been generalized to the inhomogeneously broadened case [4.22] if one assumes a Lorentzian distribution for the atomic frequency,

$$\frac{\Delta\omega}{\pi}\frac{1}{(\omega - \omega_A)^2 + \Delta\omega^2} \quad , \tag{4.74}$$

where $\Delta\omega \equiv (T_2^*)^{-1}$ is the inhomogeneous linewidth. Correspondingly, (4.73) is generalized as follows:

$$Y = I_T\{[1 + \chi_1(I_T)]^2 + [\theta - \chi_2(I_T)]^2\} \quad , \tag{4.75}$$

where χ_1 and χ_2 are the real and imaginary parts of the complex susceptibility,

$$\chi_1(I_T) = 2C\,\frac{\sigma + \sqrt{1 + I_T}}{\sqrt{1 + I_T}}\,\frac{1}{\Delta^2 + (\sigma + \sqrt{1 + I_T})^2} \quad ,$$

$$\chi_2(I_T) = \frac{2C\Delta}{\Delta^2 + (\sigma + \sqrt{1 + I_T})^2} \quad ; \quad \sigma = \frac{\Delta\omega}{\gamma} \quad . \tag{4.76}$$

For $\sigma = \Delta = \theta = 0$ (4.75) with (4.76) reduces to the state equation for absorptive bistability (4.12). On the other hand, when Δ is large enough $\chi_1(I_T)$ is negligible and one has the situation of purely dispersive bistability,

$$Y = I_T\{1 + [\theta - \chi_2(I_T)]^2\} \quad . \tag{4.77}$$

The mechanism which is at the basis of dispersive bistability has been clearly stated in [4.3], and can be summarized as follows. The frequency ω_c of the empty cavity is detuned from the frequency ω_0 of the external field, so that the empty cavity transmits only partially $[Y = I_T(1 + \theta^2)]$. On the other hand, in the case of a filled cavity the frequency of the cavity is renormalized by the interaction with the atoms, and under suitable conditions it can be taken to coincide with ω_0 so that the cavity becomes transparent. This mechanism plays a role analogous to the "bleaching" of the absorber in absorptive OB. More explicitly, the renormalized frequency is given by

$$\omega_c' = \omega_c - k\chi_2(I_T) \quad . \tag{4.78}$$

Under suitable conditions there is a value $\bar{I}_T$ such that

$$\theta - \chi_2(\bar{I}_T) = 0 \quad , \tag{4.79}$$

which by (4.72,77) means that

$$\omega_c' = \omega_0 \quad \text{and} \quad Y = I_T \quad . \tag{4.80}$$

Let us consider in detail the case of homogeneously broadened system ($\sigma = 0$). In the purely dispersive situation (4.73) reduces to

$$Y = I_T\left[1 + \left(\theta - \frac{2C\Delta}{\Delta^2 + I_T}\right)^2\right] \quad . \tag{4.81}$$

In this case, the value $\bar{I}_T$ exists provided $2C > \theta\Delta$ and is given by

$$\bar{I}_T = \Delta^2\left(\frac{2C}{\theta\Delta} - 1\right) \quad . \tag{4.82}$$

Comparison of (4.73) and (4.81) shows that (4.81) holds when $\Delta^2 \gg 1$, $\Delta\theta \gg 1$, and $\chi_1(\bar{I}_T) \ll 1$, which implies $\theta \ll \Delta$.

A sketchy analysis of (4.73) is given in [4.48]; a complete treatment can be found in [4.50]. The function $Y(I_T)$ always has a single inflection point at

$$I_{T\,\mathrm{inf}} = \frac{2C - \Delta\theta + 1}{C + \Delta\theta - 1}(\Delta^2 + 1) \quad . \tag{4.83}$$

One obtains bistability if and only if the following two conditions are satisfied:

$$I_{T\,\mathrm{inf}} > 0 \quad , \quad \left(\frac{dY}{dI_T}\right)_{I_{T\,\mathrm{inf}}} < 0 \quad . \tag{4.84}$$

From now on we shall assume that $\Delta\theta > 0$ because the case $\Delta\theta < 0$ is not interesting. Hence condition $I_{T\,\mathrm{inf}} > 0$ reads

$$2C > \Delta\theta - 1 \quad . \tag{4.85}$$

In the purely dispersive case one has $\Delta\theta \gg 1$, so that (4.85) guarantees the existence of the value $\bar{I}_T$ [cf. (4.82)]. This gives the physical interpretation of condition (4.85). On the other hand, condition $(dY/dI_T)_{I_{T\,\mathrm{inf}}} < 0$ reads explicitly

$$F(C,\Delta,\theta) \overset{\mathrm{def}}{=} (2C - \Delta\theta + 1)^2(C + 4\Delta\theta - 4) - 27C(\Delta + \theta)^2 > 0 \quad . \tag{4.86}$$

When (4.85) is satisfied and $F > 0$, one gets a hysteresis cycle for I_T vs Y; roughly speaking, the larger F is, the larger the cycle.

In Sect.4.2.1b we have seen that in the purely absorptive case ($\Delta = \theta = 0$) one gets bistability for $C > 4$. The natural question which arises is whether by taking Δ, $\theta \neq 0$ one can also obtain bistability for $C < 4$ or not. The answer is negative; for $C < 4$ not only absorptive, but also dispersive bistability is impossible. Furthermore, one finds that for $C > 4$ the cycle is largest for $\Delta = \theta = 0$. In this sense, in the case of homogeneously broadened systems absorptive OB is more convenient than dispersive OB. This is no longer true in the case of inhomogeneously broadened systems ($\sigma \neq 0$). For fixed Δ, θ, and σ one obtains bistability provided C is larger than a suitable value C_{min} which depends on Δ, θ, σ. C_{min} increases rapidly with σ. The important point is that for $\sigma \gg 1$ one finds values of C such that the system is not bistable for $\Delta = \theta = 0$ but becomes bistable for large enough Δ and θ [4.47,48]. *In other words, for these values of σ and C one does not find absorptive bistability, but only dispersive bistability. As we have seen before, this situation never occurs in homogeneously broadened systems.* Hence when inhomogeneous broadening is dominant, dispersive OB is actually easier than absorptive OB.

References

4.1 A. Szöke, V. Daneu, J. Goldhar, N.A. Kurnit: Appl. Phys. Lett. *15*, 376 (1969); J.W. Austin, L.G. Deshazer: J. Opt. Soc. Am. *61*, 650 (1971); E. Spiller: J. Appl. Phys. *43*, 1673 (1972); H. Seidel: US Patent 3,610,731 (1971)
4.2 S.L. McCall: Phys. Rev. A*9*, 1515 (1974)
4.3 H.M. Gibbs, S.L. McCall, T.N.C. Venkatesan: Phys. Rev. Lett. *36*, 113 (1976)
4.4 T.N.C. Venkatesan, S.L. McCall: Appl. Phys. Lett. *30*, 282 (1977)
4.5 F.S. Felber, J.H. Marburger: Appl. Phys. Lett. *28*, 731 (1976); Phys. Rev. A*17*, 335 (1978)
4.6 R. Bonifacio, L.A. Lugiato: Opt. Commun. *19*, 172 (1976)
4.7 P.W. Smith, E.H. Turner: Appl. Phys. Lett. *30*, 280 (1977)
4.8 T. Bishofberger, Y.R. Shen: Appl. Phys. Lett. *32*, 156 (1978); Phys. Rev. A*19*, 1169 (1979)
4.9 D. Grischkowski: J. Opt. Soc. Am. *68*, 641 (1978)
4.10 E. Garmire, J.H. Marburger, S.D. Allen, H.G. Winful: Appl. Phys. Lett. *34*, 374 (1979)
4.11 H.M. Gibbs, S.L. McCall, T.N.C. Venkatesan, A.C. Gossard, A. Passner, W. Wiegmann: Appl. Phys. Lett. *35*, 451 (1979)
4.12 D.A.B. Miller, S.D. Smith, A. Johnston: Appl. Phys. Lett. *35*, 658 (1979); Opt. Commun. *31*, 101 (1979)
4.13 W.J. Sandle, A. Gallagher: In [4.57]
4.14 G. Grynberg, E. Giacobino, M. Devaud, F. Biraben: Phys. Rev. Lett. *45*, 434 (1980)
4.15 E. Arimondo, A. Gozzini, L. Lovitch, E. Pistelli: In [4.57]
4.16 R. Bonifacio, L.A. Lugiato: Lett. Nuovo Cimento *21*, 505 (1978)
4.17 R. Bonifacio, L.A. Lugiato: Lett. Nuovo Cimento *21*, 510 (1978)
4.18 R. Bonifacio, M. Gronchi, L.A. Lugiato: Opt. Commun. *30*, 129 (1979)
4.19 R. Bonifacio, L.A. Lugiato, M. Gronchi: In *Laser Spectroscopy IV*, Proc. 4th Int. Conf., Rottach-Egern, Fed. Rep. Germany, June 11-15, 1979, ed. by H. Walther, K.W. Rothe, Springer Series in Optical Sciences, Vol.21 (Springer Berlin, Heidelberg, New York 1979)
4.20 K. Ikeda: Opt. Commun. *30*, 257 (1979); K. Ikeda, H. Daido, D. Akimoto: Phys. Rev. Lett. *45*, 709 (1980)
4.21 R. Roy, M.S. Zubairy: Phys. Rev. A*21*, 274 (1980)
4.22 M. Gronchi, L.A. Lugiato: Opt. Lett. *5*, 108 (1980)
4.23 P. Meystre: Opt. Commun. *26*, 277 (1978)
4.24 E. Abraham, R.K. Bullough, S.S. Hassan: Opt. Commun. *29*, 109 (1979); *33*, 93 (1980); *35*, 291 (1980)
4.25 H.J. Carmichael: Opt. Acta *27*, 147 (1980)
4.26 J.A. Hermann: Opt. Acta *27*, 159 (1980)
4.27 R. Roy, M.S. Zubairy: Opt. Commun. *32*, 163 (1980)
4.28 H.J. Carmichael, J.A. Hermann: Z. Phys. *38*, 365 (1980)

4.29 R. Bonifacio, L.A. Lugiato: In *Coherence and Quantum Optics IV*, Proc. 4th Conf. Rochester, USA, June 8-10, 1977, ed. by L. Mandel, F. Wolf (Plenum, New York 1978)
4.30 R. Bonifacio, L.A. Lugiato: Phys. Rev. A*18*, 1129 (1978)
4.31 R. Bonifacio, P. Meystre: Opt. Commun. *27*, 147 (1978); *29*, 131 (1978)
4.32 P. Meystre, F. Hopf: Opt. Commun. *29*, 235 (1979)
4.33 V. Benza, L.A. Lugiato: Lett. Nuovo Cimento *26*, 235 (1979)
4.34 F. Hopf, P. Meystre, P.D. Drummond, D.F. Walls: Opt. Commun. *31*, 245 (1979)
4.35 R. Bonifacio, L.A. Lugiato: Phys. Rev. Lett. *40*, 1023, 1538 (1978)
4.36 R. Bonifacio, M. Gronchi, L.A. Lugiato: Phys. Rev. A*18*, 2266 (1978)
4.37 C.R. Willis: Opt. Commun. *26*, 62 (1978)
4.38 A. Schenzle, H. Brandt: Opt. Commun. *27*, 85 (1978); *31*, 401 (1979)
4.39 F.T. Arecchi, A. Politi: Opt. Commun. *29*, 361 (1979)
4.40 R.F. Gragg, W.C. Schieve, A.R. Bulsara: Phys. Lett. A*68*, 294 (1978); Phys. Rev. A*19*, 2052 (1979)
4.41 P.D. Drummond, D.F. Walls: J. Phys. A*13*, 725 (1980); In [4.57]
4.42 G.S. Agarwal, L.M. Narducci, R. Gilmore, D.H. Feng: Opt. Lett. *2*, 88 (1978)
4.43 G.S. Agarwal, L.M. Narducci, R. Gilmore, D.H. Feng: Phys. Rev. A*18*, 620 (1978)
4.44 L.A. Lugiato: Nuovo Cimento B*50*, 89 (1979)
4.45 L.M. Narducci, R. Gilmore, D.H. Feng, G.S. Agarwal: Phys. Rev. A*20*, 545 (1979)
4.46 F. Casagrande, L.A. Lugiato: Nuovo Cimento B*55*, 173 (1980)
4.47 R. Bonifacio, L.A. Lugiato: Lett. Nuovo Cimento *21*, 517 (1978)
4.48 S.S. Hassan, P.D. Drummond, D.F. Walls: Opt. Commun. *27*, 480 (1978)
4.49 G.P. Agrawal, H.J. Carmichael: Phys. Rev. A*19*, 2074 (1979)
4.50 R. Bonifacio, M. Gronchi, L.A. Lugiato: Nuovo Cimento B*53*, 311 (1979)
4.51 P. Schwendimann: J. Phys. A*12*, L 39 (1979)
4.52 C.M. Bowden, C.C. Sung: Phys. Rev. A*19*, 2392 (1979)
4.53 C.R. Willis, J. Day: Opt. Commun. *28*, 137 (1979)
4.54 S.P. Tewari: Opt. Acta *26*, 145 (1979)
4.55 F.T. Arecchi, A. Politi: Lett. Nuovo Cimento *23*, 65 (1978)
4.56 G.P. Agrawal, C. Flytzanis: Phys. Rev. Lett. *44*, 1058 (1980); J.A. Hermann: In [4.57]
4.57 C.R. Bowden, M. Ciftan, H.R. Robl (eds.): Proceedings of the International Conference on Optical Bistability, Asheville, USA, June 3-5, 1981 (Plenum New York)
4.58 M.B. Spencer, W.E. Lamb, Jr.: Phys. Rev. A*5*, 884 (1972)
4.59 W.J. Sandle, R.J. Ballagh, A. Gallagher: In [4.57]
4.60 R. Bonifacio, P. Schwendimann : Lett. Nuovo Cimento *3*, 509,512 (1970); R. Bonifacio, P. Schwendimann, F. Haake: Phys. Rev. A*4*, 302, 854 (1971); R. Bonifacio, L.A. Lugiato: Phys. Rev. A*11*, 1507 (1975); "Atomic Cooperation in Quantum Optics: Superfluorescence and Optical Bistability" in *Pattern Formation by Dynamic Systems and Pattern Recognition*, ed. by H. Haken, Springer Series in Synergetics, Vol.5 (Springer Berlin, Heidelberg, New York 1979)
4.61 R. Landauer, J.W.F. Woo: *Synergetics*, ed. by H. Haken (Teubner, Stuttgart 1973)
4.62 L.A. Lugiato: Lett. Nuovo Cimento *23*, 609 (1978)
4.63 H. Risken, K. Nummedal: J. Appl. Phys. *49*, 4662 (1968)
4.64 R. Graham, H. Haken: Z. Phys. *213*, 420 (1968)
4.65 B.R. Mollow: Phys. Rev. A*5*, 1522 (1972); M. Sargent III: Phys. Rep. *43*, 223 (1978)
4.66 M. Gronchi, V. Benza, L.A. Lugiato, P. Meystre, M. Sangent III: Phys. Rev. A *24*, 1419 (1981)
4.67 V. Benza, L.A. Lugiato: Z. Phys. B*35*, 383 (1979)
4.68 V. Benza, L.A. Lugiato, P. Meystre: Opt. Commun. *33*, 113 (1980)
4.69 V. Benza, L.A. Lugiato: In [4.57]
4.70 H. Haken: *Synergetics – An Introduction*, 2nd ed., Springer Series in Synergetics, Vol.1 (Springer Berlin, Heidelberg, New York 1978)
4.71 H. Haken: Z. Phys. B*21*, 105 (1975); B*22*, 69 (1975)
4.72 H. Haken, H. Ohno: Opt. Commun. *16*, 205 (1976)
4.73 L.A. Lugiato: Opt. Commun. *33*, 108 (1980)
4.74 F. Casagrande, L.A. Lugiato, M.L. Asquini: Opt. Commun. *32*, 492 (1980)

4.75 M. Sargent III: Kvant Elektron. (Moscow) *10*, 2151 (1980); Sov. J. Quantum Electron. *10*, 1247 (1980)
4.76 H. Haken: "Laser Theory", in *Light and Matter Ic*, ed. by L. Genzel, Encyclopedia of Physics, Vol.XXV/2c (Springer Berlin, Heidelberg, New York 1970); W. Weidlich, F. Haake: Z. Phys. *185*, 30 (1965)
4.77 M.O. Scully, W.E. Lamb, Jr.: Phys. Rev. *159*, 208 (1967)
4.78 L.A. Lugiato: Z. Phys. B *41*, 85 (1981)
4.79 M. Gronchi, L.A. Lugiato: Lett. Nuovo Cimento *23*, 593 (1978)
4.80 B.R. Mollow: Phys. Rev. *188*, 1969 (1969)
4.81 H.J. Carmichael, D.F. Walls: J. Phys. B*9*, 1199 (1976)
4.82 H.J. Carmichael: to be published
4.83 L.A. Lugiato: Lett. Nuovo Cimento *29*, 375 (1980)
4.84 J.W.F. Woo, R. Landauer: IEEE J. Quantum Electron. *7*, 435 (1971)
4.85 A.P. Kasantsev, G.I. Surdutovich: Sov. Phys. JETP *31*, 133 (1970)
4.86 R. Salomaa, S. Stenholn: Phys. Rev. A*8*, 2695 (1973)
4.87 J.F. Scott, M. Sargent III, C. Cantrell: Opt. Commun. *15*, 13 (1975)
4.88 L.A. Lugiato, P. Mandel, S.T. Dembinski, A. Kossakowski: Phys. Rev. A*18*, 238 (1978)
4.89 A. Baczynski, A. Kossakowski, T. Marszalek: Z. Physik B*23*, 205 (1976)
4.90 R.B. Shaefer, C.R. Willis: Phys. Rev. A*13*, 1874 (1976)
4.91 L. Mandel, R. Roy, S. Singh: In [4.57]
4.92 V. Degiorgio M.O. Scully: Phys. Rev. A*2*, 1170 (1970)
4.93 R. Graham, H. Haken: Z. Phys. *237*, 31 (1970)
4.94 H.A. Kramers: Physica *7*, 284 (1940)
4.95 R. Bonifacio, L.A. Lugiato, J.D. Farina, L.M. Narducci: IEEE J. Quantum Electron QE-*17*, 357 (1981)
4.96 L.A. Lugiato: Physica A*81*, 565 (1975)
4.97 H. Risken: Z. Phys. *186*, 85 (1965); *191*, 302 (1966)
4.98 G. Nicolis, R. Lefever: Phys. Lett. A*62*, 469 (1977)
4.99 L.A. Lugiato, J.D. Farina, L.M. Narducci: Phys. Rev. A*22*, 253 (1980); J.D. Farina, L.M. Narducci, J.M. Yuan, L.A. Lugiato: Opt. Eng. *19*, 469 (1980)

5. Optical Bistability

S. L. McCall and H. M. Gibbs

Intrinsic optical bistability is reviewed with emphasis on models and experimental findings. Simple arguments are given to describe absorptive and dispersive bistability. Dispersive bistability is treated in detail. Experimental results are discussed indicating the essentials learned from particular experiments. Future prospects are contemplated.

5.1 Background

5.1.1 Early Work on Absorptive Optical Bistability

This paper is concerned primarily with intrinsic optical bistability in which the essential feedback occurs through light-matter interactions within the cavity. Systems having an amplifier as part of the intracavity medium are not treated, so the rather large literature on optical bistability in lasers is explicitly excluded. In what follows, the word bistability shall refer to intrinsic optical bistability. Consequently, it will be made clear if a system is hybrid, i.e., involves electrical feedback.

The first studies of bistability were by SZÖKE et al. [5.1] and by SEIDEL [5.2]. The essentials of SZÖKE et al.'s argument were as follows. Consider a Fabry-Perot cavity tuned to resonance with an input laser beam and which contains an absorber of optical depth αL. At low light intensities, constructive interference inside the Fabry-Perot cavity does not occur because of the absorption. The light intensity inside the cavity is $I_c \approx I_0 T$, where I_0 is the incident light intensity, and $T = 1 - R$ is either mirror's reflectivity. The transmitted light intensity I_T is therefore approximately

$$I_T = I_0(e^{-\alpha L})T^2 \quad .$$

This argument holds as long as $I_s > I_0 T$, where I_s is the saturation intensity of the absorber.

Alternatively, if the input light intensity is high, then the absorber is bleached, constructive interference occurs, $I_T \approx I_0$, and the cavity light intensity $I_c \approx I_0/T$. This argument holds as long as $I_s < I_0/T$.

The two conditions $I_s > I_0T$ and $I_s < I_0/T$ allow a range of input intensities where either argument holds (e.g., both inequalities hold when $I_0 = I_s$). The transmission should exhibit bistable behavior.

This argument is qualitative. It fails in the case of large inhomogeneous broadenings. SZÖKE et al. [5.1] considered the limit of small-mirror transmission and optical depth, and neglected standing-wave effects to find

$$I_0 = I_T\left[1 + \frac{\alpha L}{T}(1 + I_T/TI_s)^{-1}\right]^2$$

where the absorber is assumed to saturate as indicated, which follows straightforwardly from a simple two-level system which obeys rate equations, or from application of Bloch's equations to a homogeneously broadened two-level system. The condition for bistability is that $dI_0/dI_T < 0$ in some region, so that $I_T(I_0)$ is an s-shaped curve. They showed that this condition is equivalent to $\alpha L/T > 8$. Dropping the requirements that T and αL be small, they compared numerical results with the analytic result and found approximate agreement.

With SF_6 inside a Fabry-Perot cavity, and using a CO_2 laser, they observed nonlinear transmission effects, but failed to observe bistability. In retospect, we can say that Doppler broadening may have prevented bistability.

Subsequently, AUSTIN and DESHAZER [5.3], and SPILLER [5.4] investigated the transmission of cavity-enclosed absorbing dyes, but failed to see bistability. They made a number of numerical simulations including the effects of unsaturable losses.

MCCALL [5.5] calculated numerically the nonlinear and bistable transmission of a two-level medium inside a Fabry-Perot cavity, taking standing-wave effects fully into account.

All of these efforts were involved with purely absorptive bistability, i.e., any effects due to a change in medium refractive index were ignored. Furthermore, the model calculations, even MCCALL's [5.5] which included standing-wave effects, did *not* prove that bistability should occur, even with an ideal two-level system, because transverse effects were not considered. For example, with an input Gaussian beam, one could imagine that less intense parts of the beam, below the "switch-down" intensity, could through diffraction effects "turn off" any parts of the beam in the bistable region. Were that true, then only nonlinear transmission would be observed, albeit interesting.

5.1.2 First Observation of Optical Bistability; Discovery of Dispersive Optical Bistability

GIBBS et al. [5.6] observed bistability using the D lines of Na vapor. They knew that inhomogeneous or Doppler broadening was detrimental to absorptive bistability, but were unaware of dispersive bistability. Consequently, they planned to introduce a buffer gas so that a given Na atom would experience several velocity-changing collisions during a radiative lifetime, thus in effect, for steady-state bistability results, making the D lines homogeneous. Estimates of the required power were made. At a point in laser development, they decided to try the experiment even though supposedly insufficient power was available. Strong nonlinear transmission characteristics were observed, and were more pronounced as the buffer gas pressure was decreased. At zero buffer gas pressure, bistability was observed.

Bistability occurred in those experiments because the refractive index of Na vapor is nonlinear. Only a small amount of power is required to "burn" a hole in the Doppler-broadened line, through the process of pumping atoms from the F = 1 to F = 2 ground states or vice versa. At greater light intensities, atoms further removed from resonance are optically pumped, and the subsequent change in absorption line profile changes the refractive index at the laser frequency.

Such optical bistability is termed dispersive, and may be most easily understood in a model wherein the change in refractive index is proportional to light intensity, i.e., in terms of an AC Kerr effect [5.6,7]. A dependence of intracavity medium refractive index on intracavity light intensity implies a dependence of cavity resonance frequency on the intracavity light intensity. We may use an argument previously given for absorptive bistability.

Suppose the mirror spacing is such that, at low light intensities, the input laser frequency is midway between two cavity resonances. Let the input light intensity be small. The transmission of the Fabry-Perot cavity is then about T^2 [actually $T^2(4R + T^2)^{-1}$], and the intracavity intensity I_c is about $I_c = TI_0$. This argument holds as long as $TI_0 < I_r$, the cavity intensity required to change the intracavity medium's refractive index enough to shift a cavity resonance to coincide with the input laser frequency.

Suppose the input light intensity is high, but not too high. The transmission is about one, so that the intracavity intensity is about I_0T^{-1}. This argument holds as long as $I_0T^{-1} \approx I_r$. As before, there is a range of input intensities where either estimate applies, and the system should exhibit bistable behavior.

Later, it will be seen that any nonconstant dependence of refractive index on light intensity of a nonabsorbing intracavity medium allows in principle the construction of a bistable device if the cavity finesse can be made large enough.

In general, if a black box has a transmission $\mathcal{T}$ which depends on the output intensity I_T, and not necessarily on the input I_0, then the transmitted intensity is given by

$$I_T = \mathcal{T}(I_T)I_0 \quad .$$

For bistability, one requires $dI_0/dI_T < 0$ in some region, which reduces to

$$\frac{d\mathcal{T}}{dI_T} > \frac{\mathcal{T}}{I_T} \quad ,$$

i.e., a graph of $T(I_T)$ has regions where the tangent dT/dI_T is steeper than the ray T/I_T. This criterium[1] is very general and is important specifically for mirrorless hybrid bistable optical devices.

In any case, there is feedback of some sort, optical or electronic, so that T is a function of I_T. It may be that T is a function of both I_0 and I_T. Interesting additional effects then occur [5.8].

5.1.3 Hybrid Optical Bistability

The first proposal for a hybrid bistable device was by KASTAL'SKII [5.9]. The first hybrid device was constructed by SMITH and TURNER [5.10]. Part of the output intensity was detected, and the electrical signal was then amplified and used with bias added to drive an intracavity phase shifter, thus electronically simulating an intrinsic bistable device. By using a large number of Si cells in series and a large resistor in parallel with an intracavity phase shifter, an "integrated" hybrid device, i.e., one without external power, was constructed [5.11]. About the same time, GARMIRE et al. [5.12] constructed hybrid devices which had no mirrors.

5.2 Models of Optical Bistability in a Fabry-Perot Cavity

Numerous theoretical works have appeared treating bistability using various models and assumptions. Some of these treatments are covered elsewhere in this book (Chap.4).

Historically, absorptive bistability was first considered theoretically, even though it has rarely made an appearance in the laboratory[2]. Only after dispersive bistability was discovered [5.6] were dispersive and mixed absorptive and dispersive bistability considered.

We shall proceed in a general fashion to describe mixed bistability, and then isolate cases of particular interest. The approximations and simplifications made here are: that the light is a plane wave; the polarization of the light is fixed, and the medium does not change it; the mirrors are perfectly flat, have reflectivity R and transmissivity T, with T + R = 1; the light is incident from the right with one mirror at z = L, the other at z = 0. First boundary conditions will be treated in detail. Later, medium properties and the effects of another input (control) beam will be considered.

---

1 This construction is essentially given in [5.7].

2 Absorptive bistability was first observed in the work of [5.6].

5.2.1 Boundary Conditions

To the left of the cavity, only light travelling to the left, the transmitted light E_T, is present so that

$$E_T(z,t) = \varepsilon_T(z,t)\; e^{-ikz-i\omega t} + c.c \qquad z \leqq 0 \tag{5.1}$$

where $\varepsilon_T(z,t)$ is the transmitted light envelope. Inside the cavity, waves more in both directions so that

$$E_c(z,t) = \varepsilon_F(t,z)\; e^{-ikz-i\omega t} + \varepsilon_B(t,z)\; e^{+ikz-i\omega t} + c.c. \qquad 0 \leqq z \leqq L \tag{5.2}$$

where the cavity field E_c has two envelopes, the envelope ε_F for the forward field, and the envelope ε_B for the backward field. To the right of the entrance mirror the field consists of an incident wave and a reflected wave,

$$E = \varepsilon_I\; e^{-ikz-i\omega t} + \varepsilon_R\; e^{+ikz-i\omega t} + c.c. \qquad z \geqq L \tag{5.3}$$

where ε_I is the envelope for the incident field and ε_R the envelope for the reflected field.

At z = 0 a choice of phase is available, so we choose boundary conditions

$$\varepsilon_F(0,t) = \frac{1}{\sqrt{T}}\,\varepsilon_T(0,t)\;;\quad \varepsilon_B(0,t) = \sqrt{R}\,\varepsilon_F(0,t)\quad . \tag{5.4}$$

At the entrance mirror phases are important, and we may note that the boundary conditions consist of a matrix relating the outgoing fields ε_R and ε_F with the incoming fields ε_I and ε_B. The mirror is assumed to be linear, so the matrix relation is linear. Since the mirror is lossless, the matrix is unitary. Furthermore, if, for example, $\varepsilon_B = 0$, then the reflectivity R specifies an absolute value of one of the matrix elements. We may therefore write

$$\begin{pmatrix}\varepsilon_F\\ \\ \varepsilon_R\end{pmatrix} = \begin{pmatrix}\sqrt{T}\,e^{i\alpha} & \sqrt{R}\,e^{i\beta}\\ \\ -\sqrt{R}\,e^{i\gamma} & \sqrt{T}\,e^{i\delta}\end{pmatrix}\begin{pmatrix}\varepsilon_I\\ \\ \varepsilon_B\end{pmatrix} \tag{5.5}$$

at z = L, where α, β, γ, and δ are real. The minus sign prefacing one element is anticipatory. Unitarity demands $\alpha + \beta = \beta + \gamma (\mathrm{mod} 2\pi)$. We now assume that the intensity-independent part of the real part of the refractive index of the intracavity medium has effects all included in k of (5.1-3). Then in steady state, with all time derivatives zero, ε_F and ε_B are independent of z. The choice $\alpha = \beta = \gamma = \delta = 0$ then yields $\varepsilon_I = \varepsilon_T$, i.e., the mirrors are adjusted for 100% transmission in the "empty" cavity case at frequency ω. The mirrors are then an integral number of one-half wavelengths apart.

If the entrance mirror is now moved a distance less than one-half wavelength, the diagonal elements will not change, since α, β, γ, δ are clearly indpendent of R, and in the limit $R \to 0$ the mirror movement changes nothing physical. We therefore set $\alpha = \delta = 0$. For us, consequently, the most general entrance mirror boundary condition is

$$\begin{pmatrix} \varepsilon_F \\ \varepsilon_R \end{pmatrix} = \begin{pmatrix} \sqrt{T} & \sqrt{R}\, e^{i\beta} \\ -\sqrt{R}\, e^{-i\beta} & \sqrt{T} \end{pmatrix} \begin{pmatrix} \varepsilon_I \\ \varepsilon_B \end{pmatrix} \tag{5.6}$$

where β is called the detuning parameter.

One may ask how four parameters α, β, γ, δ with one constraint $\alpha + \delta = \beta + \gamma$ (mod 2π) ended up as one parameter. Implicit in the condition $\alpha = \beta = \gamma = \delta = 0$ are two conventions regarding the phase of ε_R and ε_I. If we had chosen other conditions, allowing 100% transmission, we would have still found $|\varepsilon_I| = |\varepsilon_T|$. Furthermore, the definition of the phase of ε_R is not here important.

In general, one has an optical cavity with boundary conditions. The plane-parallel Fabry-Perot cavity is only a special case used for illustrative purposes.

5.2.2 Nonlinear Medium

In addition to boundary conditions (5.4,6), one needs the medium properties to relate ε_F and ε_B at $z = 0$ to ε_F and ε_B at $z = L$. Then specifying ε_T determines ε_F and ε_B at $z = L$, whose values determine ε_I (and ε_R). Then ε_I is determined as a function of ε_T. Notice that one cannot uniquely determine ε_T as a function of ε_I, e.g., the system may be bistable. Alternatively, ε_R could be used instead of ε_T for determining ε_I.

How ε_F and ε_B depend on the medium properties is in principle measurable by illuminating the medium without mirrors. Thus far, however, estimates using a model of the medium have been used to predict when bistability should occur. Such models are, of course, ultimately based on experiment, such as of the one-way nonlinear transmission properties of a medium. First, a somewhat general procedure including standing waves but not transverse effects will be described, and then applications made to special cases.

For Maxwell's equation $(\nabla^2 - \partial^2/c^2\partial t^2)E = \frac{4\pi}{c^2}\, \partial^2 P/\partial t^2$ define P_F and P_B as the slowly varying envelope functions in

$$P(z,t) = P_F(z,t)\, e^{-ikz-i\omega t} + P_B(z,t)\, e^{+ikz-i\omega t} + \text{c.c.} \tag{5.7}$$

plus terms harmonic or constant in kz and/or ωt

to find

$$(-\partial_z + c^{-1}\partial_t)\varepsilon_F(z,t) = -2\pi i k P_F(z,t) \tag{5.8a}$$

$$(+\partial z + c^{-1}\partial_t)\varepsilon_B(z,t) = -2\pi i k P_B(z,t) \tag{5.8b}$$

using the slowly varying envelope approximation. In the general case, we expect P_F and P_B to be odd functionals of ε_F and ε_B, but apart from that it is difficult to proceed. For the moment, consider the steady-state case. Then we may define $\chi(0)$ and $\chi(2k)$ by

$$P_F(z) = \chi(0)\varepsilon_F(z) + \chi(2k)\varepsilon_B(z) \tag{5.9a}$$

$$P_B(z) = \chi(0)\varepsilon_B(z) + \chi(-2k)\varepsilon_F(z) \quad , \tag{5.9b}$$

the form of the first equation following from the oddness property, and the second from the first and some assumptions of symmetry. The $\chi(0)$ and $\chi(2k)$ are even functions of ε_F and ε_B. As a specific example, consider the case where

$$P(z,t) = \frac{\chi E}{1 + \langle E^2\rangle/E_s^2} \tag{5.10}$$

where χ is a constant, and $\langle\rangle$ denotes a time average. This result follows from Bloch's equations, for example. Implicit is that a local response applies, i.e., the polarization at z depends on the field at z, but not on the field at any point near but not at z. Diffusion of excitation would violate this statement.

Given this specific form, we may then proceed to find P_F and P_B by averaging over a wavelength, and substitute into (5.8) with $\partial_t\varepsilon_F = \partial_t\varepsilon_B = 0$, thus relating the fields at z = L to the fields at z = 0. Boundary conditions then yield a relationship between E_T and E_I, which may be bistable. This and other models will be used in the following.

5.2.3 Conditions for Dispersive Bistability

By dispersive bistability is meant that the intracavity medium's absorption can be negelected. We shall find that the requirements for dispersive bistability are essentially that the change in refractive index is large enough to shift the Fabry-Perot cavity resonance by about one instrument function. Types of nonlinear refractive indices include the AC Kerr effect as in nitrobenzene and nonlinear refractive indices due to weak absorption such as in Na vapor, ruby, GaAs, and the thermal devices.

There is no need to be specific as to the particular origin of refractive index change. For the nonlinear part, we may write for the change in susceptibility at a point

$$\chi = \chi(|\varepsilon_F\, e^{-ikz} + \varepsilon_B\, e^{ikz}|^2) \quad ,$$

and expand in a Fourier series to find, with $\phi = -2\,kz$,

$$\chi(0) = (2\pi)^{-1}\int_0^{2\pi} d\phi\,\chi(|\varepsilon_F|^2 + |\varepsilon_B|^2 + \varepsilon_F\varepsilon_B^*e^{i\phi} + \varepsilon_F^*\varepsilon_B\, e^{-i\phi}) \quad , \tag{5.11a}$$

$$\chi(\pm 2k) = (2\pi)^{-1} \int_0^{2\pi} d\phi \chi(|\varepsilon_F|^2 + |\varepsilon_B|^2 + \varepsilon_F\varepsilon_B^* e^{i\phi} + \varepsilon_F^*\varepsilon_B e^{-i\varepsilon}) e^{+i\phi} \quad . \tag{5.11b}$$

For dispersive bistability χ is real, so that $\chi(0)$ is real and $\chi(-2k) = \chi^*(2k)$. By translating the variable of integration ϕ, one sees that $\chi(2k)$ has the same modulus as $\varepsilon_F\varepsilon_B^*$, i.e.,

$$\chi(2k) = \frac{\varepsilon_F\varepsilon_B^*}{|\varepsilon_F\varepsilon_B|} \, |\chi(2k)| \quad . \tag{5.12}$$

We may derive these results even when χ is a nonlocal function of the intensity as long as the medium is homogeneous. One concludes that extrema of χ and the light intensity coincide. By performing a physical translation in position, one finds the results above.

It then immediately follows that in steady-state

$$\frac{\partial}{\partial z} |\varepsilon_F|^2 = \frac{\partial}{\partial z} |\varepsilon_B|^2 = 0 \quad .$$

Place $\varepsilon_F = |\varepsilon_F/e^{i\phi_F}$ and $\varepsilon_B = |\varepsilon_B/e^{-i\phi_B}$. Substituting, one finds

$$\frac{\partial}{\partial z}(\phi_F + \phi_B) = 2\pi k\left[2\chi(0) + |\chi(2k)| \left(|\frac{\varepsilon_F}{\varepsilon_B}| + |\frac{\varepsilon_B}{\varepsilon_F}|\right)\right] \quad , \tag{5.13}$$

which has a z-independent right-hand side. The total additional round-trip phase shift is therefore

$$\phi = \phi_F + \phi_B = 2\pi k[2\chi(0) + |\chi(2k)|(|\varepsilon_F/\varepsilon_B| + |\varepsilon_B/\varepsilon_F|)]L \quad , \tag{5.14}$$

where L is the medium length.

Boundary conditions yield

$$\varepsilon_F = \varepsilon_T T^{-1/2} \quad , \quad \varepsilon_B = R^{1/2} T^{-1/2} \varepsilon_T \tag{5.15}$$

at the exit mirror; here ε_T is taken to be real and positive. At the entrance mirror,

$$\varepsilon_F = T^{1/2}\varepsilon_I + R^{1/2} e^{i\beta}\varepsilon_B \quad , \tag{5.16}$$

so that

$$|\varepsilon_I|^2 = [1 + R^2 - 2R\cos(\phi - \beta)]|\varepsilon_T|^2/T \quad , \tag{5.17}$$

the state equation for dispersive bistability. The round-trip phase shift ϕ is a function of the cavity intensity $|\varepsilon_C|^2 = |\varepsilon_T|^2/T$, and weakly of R. For a local AC Kerr effect model,

$$\chi = a|\varepsilon_F e^{-ikz} + \varepsilon_B e^{+ikz}|^2 \quad , \tag{5.18}$$

where a is a constant, so that $\chi(0) = a(|\varepsilon_F|^2 + |\varepsilon_B|^2), \chi(2k) = a\varepsilon_F\varepsilon_B^*$. Then

$$\phi = 6\pi k a(|\varepsilon_F|^2 + |\varepsilon_B|^2) = 6\pi k\left(\frac{1+R}{T}\right)|\varepsilon_T|^2 \alpha L \quad . \tag{5.19}$$

In general, denote $P_T = |\varepsilon_T|^2$ and $P_I = |\varepsilon_I|^2$, and ϕ is a function of P_T. The bistability condition $dP_I/dP_T < 0$ yields

$$\cos(\phi - \beta) - P_T \frac{d\phi}{dP_T} \sin(\phi - \beta) > \frac{1 + R^2}{2R} \quad . \tag{5.20}$$

The mistuning parameter is usually experimentally controllable. Define μ as the first- or fourth-quadrant angle

$$\mu = \tan^{-1}\left(P_T \frac{d\phi}{dP_T}\right) \quad , \tag{5.21}$$

so that (5.21) becomes

$$\left[1 + \left(P_T \frac{d\phi}{dP_T}\right)^2\right]^{\frac{1}{2}} \cos(\mu + \phi - \beta) > \frac{1 + R^2}{2R} \quad , \tag{5.22}$$

which for any given P_T is most readily satisfied by adjusting $\beta = \phi + \mu$. Thus if the mistuning parameter is adjustable, bistability occurs if

$$\sup\left[1 + \left(P_T \frac{d\phi}{dP_T}\right)^2\right]^{\frac{1}{2}} > \frac{1 + R^2}{2R} \tag{5.23}$$

where sup means the maximum value achieved when P_T takes on all positive values. This may be expressed

$$R > \sqrt{1 + \left(\sup\left\{P_T \left|\frac{d\phi}{dP_T}\right|\right\}\right)^2} - \sup\left\{P_T \left|\frac{d\phi}{dP_T}\right|\right\} \quad . \tag{5.24}$$

The right-hand side depends on R through (5.14) and boundary conditions, but only in second order. If $P_T \, d\phi/dP_T \ll 1$, then

$$T < \sup\left\{P_T \left|\frac{d\phi}{dP_T}\right|\right\} \quad . \tag{5.25}$$

Clearly, any dependence of ϕ on P_T allows bistability for sufficiently small mirror transmission T. In the AC Kerr effect approximation, any nonzero mirror transmission T will allow bistability at some input power. If the round-trip phase shift $\phi(P_T)$ saturates, then we may estimate sup $P_T \, |d\phi/dP_T|$ as $\frac{1}{4}\phi_s$, where ϕ_s is the saturation value of ϕ. Then (5.24) reduces to

$$\phi_s > 4T \tag{5.26}$$

which is about two instrument function widths.

The required input power at R given by (5.24) as an equality at the corresponding inflection point is found to be

$$P_I = \frac{(P_T d\phi/dP_T)^2}{1 + (P_T d\phi/dP_T)^2 - \sqrt{1 + (P_T d\phi/dP_T)^2}} P_T \tag{5.27}$$

where (5.17,21,24) are used as equations, and $P_T d\phi/dP_T$ and P_T are evaluated where $|P_T d\phi/dP_T|$ is a supremium. For the case when the supremium $\ll 1$, then

$$P_I = 2P_T \tag{5.28}$$

corresponding to $T = \sup|P_T d\phi/dP_T|$.

It may be desired to minimize the value of the input required to observe bistability. In fact, the conditions that lead to (5.27) also yield the P_I conditions for minimizing P_I. In other words, any input power greater than that given by (5.27) will be in a bistable region for some R and β. The reader can be convinced of this by noting that for the detuning parameter given by $\beta = \phi - \mu$ evaluated at P_T such that $P_T|d\phi/dP_T|$ is a maximum, an increase in R beyond the boundary of the inequality (5.24) will lead to a negative slope at the point.

If the quantity $|P_T d\phi/dP_T|$ has no maximum (e.g., AC Kerr effect), then one may wish to minimize the value of P_I required for bistability.

5.3 Experimental Findings

Here, we shall review the experiments in intrinsic optical bistability, retrieving from them the essentials for future systems. The systems studied shall be Na vapor [5.6], ruby [5.13], thermal, Kerr liquid [5.14], thermal [5.15], and GaAs [5.16] systems.

5.3.1 Na Vapor

Before the Na vapor experiment was performed, it was known that absorptive bistability occurred in a plane wave model [5.1,2] including standing waves [5.5] as long as there was sufficient mirror reflectivity to convert a negative resistance feature [5.5] into a negative slope region. The first measurements were made without mirrors in order to determine whether sufficient negative atomic conductivity occurred. It was believed that, since inhomogeneous broadening greatly reduced the amount of negative conductivity, it was necessary to include a buffer gas to prevent hole burning in the Na vapor, thus "homogenizing" the Na vapor lines. Various argon and Na densities were used to find transmitted-vs-incident power curves. The most nonlinear appeared to have only a little negative conductivity.

In spite of this unwelcome development, which was ascribed to a poorly understood ground-state-hyperfine effect, mirrors were attached, and nonlinear effects were observed. At finite argon pressures, strong nonlinear effects were observed, but not bistability. It was thought that since the wings of the optical beam were always in the low-transmitting region, that perhaps diffraction forced all parts of the beam in the bistable region to spontaneously switch to the lower state. It was noted that nonlinear effects became stronger as the argon pressure was reduced. It was decided by one of the authors of that work, without the other's knowledge, to reduce the argon pressure to zero, thereby allowing Na to slowly deposit on the mirror surfaces and possibly burning the mirror coatings. Even stronger nonlinear behavior was observed and, later, bistable transmission was seen.

To stabilize the laser frequency, a saturation absorption cell with Na vapor was used to measure the laser frequency. The initial efforts were made with the laser frequency tuned to a characteristic saturation signal. When the laser frequency was changed to another frequency, but inside the Doppler line, bistability was observed.

The observed results were not symmetric in laser frequency changes or in mirror separation changes. Clearly, the device was not performing as anticipated because certain laser frequencies were not allowed, and because the Doppler profile should have reduced the negative conductivity. The asymmetry provided the clue that dispersive effects were involved. Absorptive effects are symmetric in laser tuning from line centers; dispersive effects are asymmetric.

It was known that Na vapor had a large nonlinear coefficient of refractive index. In the laser power region of this experiment, optical pumping effects move a population from the $F = 1$ to $F = 2$ (or 2 to 1) ground-state manifolds. This changes the absorption line shape, and typically changes the refractive index. It was shown that any dependence of refractive index on light intensity can lead to optical bistability. The particular case of a linear dependence on light intensity leads to a single-parameter set of curves. The experimental results could be fit by a mixture of absorptive and dispersive bistability, but that was, of course, an incomplete model.

The main two lessons that should be learned from the Na vapor experiment are that 1) transverse effects do not destroy optical bistability, and 2) dispersive optical bistability exists.

5.3.2 Ruby

The ruby experiment was designed to study the homogeneously broadened line case and to observe optical bistability in the standing-wave case. A tunable ruby laser was constructed to do this, and the device was at temperatures from near liquid nitrogen to above room temperature. It was anticipated that bistability would be observed only near liquid nitrogen temperature; otherwise estimates of required power were excessive, assuming the only contributions to a nonlinear susceptibility come from the R_1 line saturation. It was found that bistability occurred at 296 K; at this temperature the laser output was approximately midway between the R_1 and R_2 lines. Furthermore, the absorption, though finite, was small. Thus the bistability could not be absorptive. Dispersive contributions from the R_1 and R_2 lines partially cancelled, and the dispersive contribution from either was far too small in any event. The nonlinearity responsible for the bistability was dispersive, however, and understood through an off-resonance process. The slight absorption by the R_1 and R_2 lines caused population changes. The refractive-index contribution from Cr^{3+} ions was changed because higher lying states, in particular the charge transfer state in the ultraviolet, were driven from a different initial state. The ruby shows that by driving a weakly absorbing transition, thereby changing population distributions, it is possible to obtain large nonlinear-refractive-index contributions from nonresonant levels. This effect might be exploited in semiconductors with exciton or impurity-level transitions.

5.3.3 Kerr Materials

Traditionally, high-power lasers (e.g., Q-switched ruby) were used to study nonlinear-refractive-index effects in materials such as nitrobenzene or CS_2. Using a pulsed ruby laser, studies were made of optical bistability [5.14] using such optical Kerr liquids inside a Fabry-Perot. Models and experiment agreed. It was found that the switch-on and switch-off times were explained only if one took into account the cavity-ringing time. When the cavity-ringing time is long compared to the medium relaxation time, the device switches on or off at a time delayed from when the input crosses a switching intensity, the delay depending on the rate of change of input. When the output has made a large fractional change towards the other state, the switching is characterized by a few oscillations which damp in a cavity-ringing time. In micron-size devices the cavity-ringing time will be short, and materials times will probably dominate. The delay is due to the fact that at the switching point all characteristic rates are zero, and a finite excursion from equilibrium is required for exponential runaway. The "critical slowing-down phenomenon" is avoided when one switches with pulses with energies about twice threshold.

5.3.4 Thermal Bistability

Thermal optical bistability [5.15] was first seen in GaAs, and studied with absorbing glass as the medium. Using 57-μ-thick heatsunk glass and a 50-μ-diameter laser beam, turn-on and turn-off time-domain measurements were made. The heat-diffusion equation for a plate heatsunk on one side is solvable using eigenfunctions and eigenvalues. The results were modelled using a single eigenvalue corresponding to the longest lifetime with good agreement. The switch-on time development is the only case where clear disagreement occurred, and is ascribed to the fact that transverse effects are important in heat transport on a short time scale. With time scales easily adjustable, such systems can be useful for studying one's ability to model. Furthermore, when distances are pushed to micron sizes, and high-thermal-conductivity materials (e.g., semiconductors) are used, the material thermal time constants become subnanosecond.

5.3.5 Semiconductors

There are a number of candidates among semiconductors for optical bistability. Unless the thickness is quite small, one would use direct bandgap materials or materials with impurity states below the gap. The reason is that indirect phonon-assisted transitions are weak and just cause background absorption. In very thin samples such absorption may not be so bad, however. Materials such as GaAs, InSb, and CdS are candidates [5.17]. Among the doped semiconductors, N-doped GaP is a representative [5.17]. The motivation for going to semiconductors is that semiconductors are about the closest thing there is to a solid-state, high-density 2-level system without huge linewidths. Actually, Na vapor can have such high densities that it would be suitable, but it seems fruitless to pursue that direction because from a practival viewpoint no one wants to deal with such a caustic material. Ideally, one wants a system which is solid state, stable, requires less than femtojoule switching energies, switches in picoseconds, has large fan-out capability, does not glitch, is reasonably fabricated, works at room temperature, etc. for potential practical applications. It is surprising how close to this ideal one can project with confidence based on present results.

Optical bistability has been observed in two semiconductors, in GaAs and somewhat later in InSb. The GaAs experiment used a molecular-beam-grown GaAlAs-GaAs-GaAlAs sandwich. At low temperatures (e.g., 100 K), the exciton feature in high-quality GaAs is quite sharp; at the lowest temperature, the low-intensity absorption feature is 7×10^4 cm^{-1} in absorption in some samples. The resonant frequency is about 1.5 eV, barely infrared, and at a popular point in the spectrum for communications. (This popular region probably will move to about 0.8 eV, however.) It has been shown experimentally that the exciton absorption feature saturates as the sum of a fairly small unsaturable background and a simple saturable absorber. The unsaturable background prevented purely absorptive bistability in that sample. In better samples, that story may be different. Bistability was observed off-resonance, however, where the unsaturable background was probably a lot smaller and where dispersive effects dominated. The mechanism for nonlinear dispersion is as follows. Light below the gap and below the exciton features creates carriers. Because the sample is at a finite temperature, the carriers quickly develop a Fermi-Dirac distribution among the states within the valence and conduction bands. The absorption is thereby changed, and decreased near the bandedge, in particular. There is some band spilling due to carrier interactions, so that the absorptivity below the exciton feature does not necessarily completely saturate. Among the absorption features the exciton feature is the first to decrease because it is the smallest energy feature of any importance, and because the electron-hole interaction is screened by free carriers and other excitons.

In InSb a similar description should apply except that, since the exciton binding energy is so small and so easily screened by impurities, excitons do not appear as a feature. The lowest lying states are excitonlike in the sense that Coulomb interactions increase their absorptivity, so that only a relatively few states near the bottom of the band need be filled to change the absorptivity significantly.

The carrier recombination time in GaAs is about ten nanoseconds. Using microsecond-length triangular-wave input intensities, switch-up and switch-down times of about 40 nanoseconds were observed. The input was almost 100 milliwatts focussed to a 10-μ circle. The mirror reflectivity was 90%. By injecting a 200-ps yellow pulse into the sample, a switching energy of 0.6 nJ was found which represents the energy absorbed by the device, not the energy controlled by the device. The 0.6 nJ energy was spread over a 50-μ circle; thus a figure of 24 pJ is given for a switching energy.

Bistability was observed at temperatures below 120 K. At near room temperature, bistability was observed, but the mechanism was due to thermal effects, easily distinguished from electronic effects by the time response and sign of refractive index change.

Publications regarding the InSb observations are not complete as those for the GaAs work [5.16]. An electronic nonlinearity of at least 0.1 cm^2/kW, has been measured compared with 0.4 cm^2/kW deduced for GaAs at the optimum frequency from nonlinear absorption data. This nonlinearity is large enough to account for the optical bistability observed in InSb. The possibility of a thermal contribution which dominates the electronic contribution in the observed bistability has not been experimentally eliminated, however, since both could occur simultaneously, as in GaAs.

5.4 Future Prospects

5.4.1 Optical Processing and Computing

There are at least two possible uses for optical bistable systems, one in communications and the other in computers. Communication systems are evolving toward the use of light pulses to carry information. Systems now envisaged only use light pulses to carry information, with processing done electronically. Particularly in repeaters, where light pulses are detected, the resultant electrical pulses are reshaped and amplified, and light pulses retransmitted in the same sequence, but also in any processing wherein the input and output are sequences of light pulses it would be convenient to perform the processing optically in order to avoid optical-electrical interfacing. In computers one advantage that optical bistable devices might have is that they may work faster than semiconductor gates at similar switching energies. Connections may be less space consuming because spaces between light guides can be smaller than between electrical conductors without pick-up problems. Also different light frequencies can possibly be used in the same lightguide. In systems which need to be secure against large electrical interference, all-optical signal processing would be at a decided advantage.

5.4.2 Theoretical Limits on Minimum Size

These potential advantages are irrelevant unless the switching energy can be made small for bistable devices. First, two theoretical limits are established, and then extrapolations from present results are made. Then comparison is made with existing and futuristic semiconductor and Josephson technology.

The first theoretical limit is based on semiclassical equations. For bistability one needs $\alpha L/T > 8$. An optical cavity can have a waist about λ^2 in area. We may choose $T = 0.05$, so $\alpha L \sim 1$ is needed. Use a single atom whose absorption line is only broadened by lifetime effects. The absorption cross section is about λ^2, so that $\alpha L \sim 1$ is achieved. Semiclassically, this device should be bistable.

The second theoretical limit is based on the first. The first will "glitch" too often. Consequently, one needs about 1000 two-level atoms, or the statistical equivalent. This amounts to a switching energy (using 1.5 eV photons) of 2.4×10^{-16} joules, about one-fourth femtojoule.

5.4.3 Approaching Theoretical Limits

Can this be achieved? We have results on GaAs [5.16] and InSb [5.15,17] devices available. The n_2 in GaAs (0.4 cm^2/kW) deduced from nonlinear absorption data is a close to the n_2 measured in InSb 1.0 cm^2/kW). Also, more information is available to us about GaAs, so the extrapolation will be based on GaAs results.

The measured switching energy in the GaAs device was 600,000 fJ over $(50\ \mu)^2$. The wavelength of light in the material is 2300 Å, so that an optical cavity could concentrate the light into an active diameter of about 1/4 μ. The switching energy would then be 15 fJ.

The active length of the GaAs device was 4.5 μ. Peak absorptivities of 7 μ^{-1} have been observed in good GaAs. The GaAs device did not have this high absorptivity. With such absorptivity, one should need only about 2/7-μ length of active material. The switching energy should then reduce to 0.95 fJ. That is about four times the theoretical limit given above.

By using lower-frequency light, the switching energy might be reduced. The light photon energy must be several tens of times the operating temperature, however. In fact, the limit of 1000 quanta switching energy evidently applies to semiconductor and Josephson devices also, yielding limits of about 1/4 fJ and about 10^{-18} joules, respectively.

Can room-temperature operation be achieved? Bistability was achieved up to 120 K. Furthermore, with use of superlattices, the exciton feature can be moved further from the band absorption edge. Lifetimes can be shortened by introducing impurities, hopefully in a way that does not significantly broaden the exciton feature. The only other limitation is the cavity-build-up time. A cavity 1 μ long with T = 0.1 will take about 0.12 ps to build up.

5.4.4 Comparison with Other Technologies

How does this compare with futuristic semiconductor technology? Gates with transit times of 10 ps seem to be the present state of art, and 1 ps may be a limit governed by electron velocity and a required thickness of about 1000 Å. These speeds must be typically multiplied by three factors of 3. One 3 is for fanout. Another is for charging lines. Another is for additional capacity unavoidably associated with the gate in use. Thus an effective gate delay of about 27 ps seems to be a limit. Other semiconductors may reduce this time by a factor of perhaps 3.

Josephson devices have been demonstrated which switch in 15 ps, and probably 1 ps is achievable. For bistable optical devices the factors of three either are not there or are already included. If switching is done not through input mirrors, bistable optical devices already have a fanout capability.

5.4.5 Quantum Aspects

Not discussed in this article are the mathematical analogies with phase transitions since they are well dealt with elsewhere. The quantum aspects could become of importance relatively soon in a potentially practical device. In a semiconductor device wherein light absorption leads to free carriers instead of excitons, if an electric field is also placed across the semiconductor, the free carriers will be accelerated and produce more carriers, as in an avalanche photodiode. As far as a refractive-index change is concerned, the quantum efficiency is then greater than unity by a factor. This reduces the light-switching energy, an advantage, but would possibly increase the energy required to be dissipated, a disadvantage. Although electrical energy is used, such a device is not really a hybrid device because the electrical limitations, such as the effects of unavoidable capacitance, are not present. The material in the optical cavity is simply a semiconductor under an applied field. The factors of three, previously mentioned, would also be absent and all potential advantages of intrinsic bistable optical devices would be present. With greater than unity quantum efficiency, the statistical effects would become very important.

5.4.6 Preferred Wavelengths

In communications, improvements in glass-fiber production have led to a highly transparent, very low-dispersion region of the spectrum in the infrared at about 1.3-μ wavelength. It seems certain that this wavelength region will be favored in the future. Semiconductors with gaps in this region are alloys, and random concentration fluctuations lead to smearing of the gap. The smearing conceivably might be eliminated if concentration fluctuations could be eliminated, but it is not known how this could be achieved. For optical computers, it may be advantageous to move into the visible or even ultraviolet part of the spectrum, so that bistable optical devices can be made even smaller than imagined above.

References

5.1 A. Szöke, V. Daneu, J. Goldhar, N.A. Kurnit: Appl. Phys. Lett. *15*, 376 (1969); A. Szöke: U.S. Patent 3,813,605 (1974)
5.2 H. Seidel: U.S. Patent 3,610,731 (1971) (filed May 19, 1969, i.e., before both entries in Ref.[5.1])
5.3 J.W. Austin, L.G. DeShazer: J. Opt. Soc. Am. *61*, 650 1971); J.W. Austin: Ph.D. Dissertation, University of Southern California (1972)
5.4 E. Spiller: J. Opt. Soc. Am. *61*, 669 (1971); J. Appl. Phys. *43*, 1673 (1972)
5.5 S.L. McCall: Phys. Rev. A*9*, 1515 (1974)
5.6 H.M. Gibbs, S.L. McCall, T.N.C. Venkatesan: Phys. Rev. Lett. *36*, 1135 (1976)
5.7 J.H. Marburger, F.S. Felber: Phys. Rev. A*17*, 335 (1978)
5.8 A. Feldman: Opt. Lett. *4*, 115 (1979); Appl. Phys. Lett. *33*, 243 (1978)
5.9 A.A. Kastal'skii: Sov. Phys. Semicond. *7*, 635 (1973)
5.10 P.W. Smith, E.H. Turner: Appl. Phys. Lett. *30*, 280 (1977)
5.11 P.S. Cross, R.V. Schmidt, R.L. Thornton, P.W. Smith: IEEE J. Quantum Electron. *14*, 577 (1978)
5.12 E. Garmire, J.H. Marburger, S.D. Allen: Appl. Phys. Lett. *32*, 320 (1978); M. Okaka, K. Takizawa: IEEE J. Quantum Electron. *15*, 82 (1979)
5.13 T. Bischofberger, Y.R. Shen: Appl. Phys. Lett. *32*, 156 (1978); Opt. Lett. *4*, 40 (1979); Phys. Rev. A*19*, 1169 (1979)
5.14 S.L. McCall, H.M. Gibbs: J. Opt. Soc. Am. *68*, 1378 (1978); S.L. McCall, H.M. Gibbs, W. Greene, A. Passner: Unpublished
5.15 H.M. Gibbs, S.L. McCall, T.N.C. Venkatesan: U.S. Patents 4, 012, 699 (1977), 4, 121, 167 (1978)
5.16 H.M. Gibbs, S.L. McCall, T.N.C. Venkatesan, A.C. Gossard, A. Passner, W. Wiegmann: In *Digest of 1979 IEEE/OSA Conference on Laser Engineering and Applications* (IEEE, New York 1979); In *Laser Spectroscopy IV*, ed. by H. Walther, K.W. Rothe, Springer Series in Optical Sciences, Vol. 21 (Springer, Berlin, Heidelberg, New York 1979); Appl. Phys. Lett. *35*, 451 (1979)
5.17 D.A.B. Miller, S.D. Smith, A. Johnston: Appl. Phys. Lett. *35*, 658 (1979); D.A.B. Miller, S.D. Smith: Opt. Commun. *31*, 101 (1979)

6. Superfluorescence Experiments

Q. H. F. Vrehen and H. M. Gibbs

With 20 Figures

Superfluorescence (SF) is the cooperative emission of a system of many two-level atoms which are all in the upper state initially. The conditions for observing SF and the techniques employed are discussed. The various experiments are surveyed and their major contributions summarized. Single-pulse observations in cesium vapor are emphasized. The discrepancy between the ringing predicted by uniform-plane-wave simulations of Maxwell-Bloch equations and the absence of ringing in the cesium data is attributed to transverse effects. The theory of the quantum initiation of SF is reviewed. Good agreement is found between quantum predictions and cesium data for the initial tipping angle and quantum fluctuations in the delay time. Effects of sample length, Fresnel number, degeneracies, beats, and polarization are also discussed. Some open questions are outlined.

6.1 Background

Superfluorescence (SF) is the cooperative emission of a system of many two-level atoms which are all in the upper state initially[1]. The latter condition implies that at t = 0 no macroscopic polarization exists in the sample, and consequently SF requires a fully quantum-mechanical description. The theory of SF has a long and interesting history dating back to the original paper by DICKE [6.2] in 1954. In the present article, which concentrates on experiments, no justice can be done to the many theoretical contributions. The reader may consult Chap.4, or the references in the recent work of POLDER et al. [6.4]. Some of the early theories considered pointlike samples, i.e., samples with all atoms contained in a volume with linear dimensions small compared to λ, the wavelength of the SF radiation. Pointlike samples, however, would require very special atomic configurations [6.5] and

1 This use of the word superfluorescence was suggested by BONIFACIO and LUGIATO [6.1]; they have specified conditions for ideal SF. The possibility of coherent emission arising from totally inverted atoms was mentioned for a small volume in DICKE's famous paper on superradiance [6.2] and was described in qualitative detail as a "coherence brightened laser" by DICKE [6.3], but detailed conditions for SF in a large volume were first spelled out in [6.1].

no experiments on such configurations have been published. Therefore the discussion will be restricted to extended samples with all dimensions large compared to λ, and in particular to samples that take the shape of a thin pencil. The SF radiation is then characterized by the emission of a delayed pulse with peak intensity proportional to N^2 within a narrow cone in both directions along the sample axis. N is the total number of atoms.

The complexity of SF in extended samples arises from the fact that propagation effects must be taken into account and that the envelopes of field and polarization depend not only on time, but also on the three coordinates of position in the sample. The quantum-mechanical mean-field theory of BONIFACIO and LUGIATO [6.1] neglects the spatial variation of the envelopes. The semiclassical calculation of MacGILLIVRAY and FELD [6.6] fully allows for propagation effects but it is one dimensional in the sense that it only allows for axial variation of the envelopes. Because of its semiclassical nature it does not describe the initiation of SF satisfactorily. Fully retarded quantum-mechanical treatments of SF have been provided recently by GLAUBER and HAAKE [6.7] and by POLDER et al. [6.4,8]. These theories are particularly interesting for their discussion of the initiation and the quantum fluctuations. Their main limitation is that they are one dimensional and neglect transverse variations of the polarization and the field.

The history of SF experiments begins with the pioneering work of Feld et al. [6.9,10] in 1973. Their observations on hydrogen fluoride verified the characteristic properties of SF and indicated the importance of propagation effects. For an unambigous test of the theory it is desirable that experiments are performed on the simplest possible system and obey a set of conditions first formulated by BONIFACIO and LUGIATO [6.1] for their regime of "pure SF", even though these conditions may be slightly too restrictive. So far only the experiments by the present authors [6.11] on atomic cesium have met those conditions. The cesium experiment has established the existence of a regime of single-pulse emission, in qualitative agreement with mean-field theory, but the observed pulse widths confirm the importance of propagation effects. The experiment has further demonstrated that transverse variations of polarization and field must be taken into account, and it has stimulated the theoretical and experimental study of the initiation and of the quantum fluctuations.

Many other experiments have been reported since 1976. The wavelength range has been extended into the visible, various pumping schemes have been investigated, and quantum beats on coupled and on independent transitions have been observed as well as cascade SF. The effect of homogeneous and inhomogeneous atomic dephasing has been studied, polarization effects resulting from level degeneracies have been described, and so have the effects of length and Fresnel number variation and the gradual transition from SF to amplified spontaneous emission (ASE). It is the purpose of this paper to review these developments. Most attention will be given to thos experiments that promote our understanding of the basic phenomenon.

6.2 Experimental Parameters

6.2.1 Conditions for Superfluorescence

Superfluorescence (SF) is the cooperative emission of many initially inverted atoms [6.1-3]. Unlike spontaneous emission, which is a single-atom process, SF depends on the shape of the sample. This is because initially there is no macroscopic polarization in the SF sample, so only the geometry defines a preferred emission direction. Usually a thin pencil geometry is selected resulting in emission from both ends on every shot. In the case of a pencil geometry prepared with a macroscopic polarization initially, the resulting superradiance (SR) [6.2] is emitted only along the direction of the excitation. Thus the SR emerges from only one end or even transverse to the pencil if the initial polarization is phased in that direction. Also, SF begins as more or less isotropic spontaneous emission proportional to the total number of inverted atoms N and evolves to cooperative emission proportional to N^2 and directed out the ends of the pencil. SR is directed cooperative emission proportional to N^2 from the very beginning.

Consequently, the first condition for SF must be complete inversion of the radiating system in order that it superfluoresce, i.e., build up from noise, rather than superradiate. The inversion is sufficiently complete if, initially, spontaneous emission dominates over cooperative emission. A nondegenerate two-level transition should be used for studying SF in order to avoid complications from sublevel interferences and competing transitions.

In order that the emission be cooperative, the characteristic time of the SF process must be shorter than the relaxation times which can destroy the cooperative interactions between radiating atoms via the axial modes of the radiation field. The length of the SF process is characterized by the delay time τ_D between the creation of the inversion and the peak of the SF emission. The relevant relaxation times are the inversion relaxation time, T_1; the single-atom transverse polarization relaxation time, T_2'; and the many-atom transverse polarization relaxation time from inhomogeneous effects, T_2^*. Since the SF described herein occurs in gases, T_1 and T_2' are usually determined by radiative spontaneous emission, foreign gas collisions, or escape from the light beam. T_2^* is usually determined by Doppler broadening or inhomogeneous external static fields.

Typically, τ_D is 10 to 100 times longer than the SF time τ_R defined by

$$\tau_R = \frac{8\pi\tau_0}{3n\lambda^2 L} \tag{6.1}$$

where τ_0 is the partial lifetime of the SF transition and is equal to or longer than the total radiative lifetime of the upper state, n is the inversion density, λ is the SF wavelength, and L is the length of the sample of the inverted atoms. τ_R is approximately the time for a spontaneously emitted photon to be emitted along the pencil. Alternatively, it is roughly the time for the initial quantum emission to become classical [6.12]. In order to avoid complications from the finite sample length, one should take τ_R longer than the sample transit or escape time, $\tau_E = L/c$. If atoms are to emit cooperatively, they must be able to communicate more quickly than they can radiate. The ARECCHI-COURTENS [6.13] cooperation time τ_c is just the value of τ_R for a sample of length equal to the distance light travels in that time τ_c,

$$\tau_c = \tau_R\Big|_{L = c\tau_c} = \sqrt{\tau_R\tau_E} \quad . \tag{6.2}$$

The excitation process should be characterized by a time τ_p much shorter than the process evolution time τ_D in order that the details of the preparation of the inversion can be neglected in the analysis of its subsequent evolution. It would be even safer to require $\tau_p < \tau_R$, but simulations show that such a stringent requirement is unnecessary.

Finally, a Fresnel number F of about unity is needed so emission occurs in a single transverse mode with small diffraction losses; $F = A/\lambda L$ where A is the cross-sectional area.

The desirable conditions for superfluorescence may be summarized as follows: inverted nondegenerate two-level atoms with sufficiently long relaxation times and appropriate sample length and inversion density to satisfy

$$\tau_E < \tau_c < \tau_R < \tau_D < T_1, T_2', T_2^* \quad . \tag{6.3}$$

The inversion should be prepared in a time much shorter than the SF build-up time, i.e.,

$$\tau_p < \tau_D \tag{6.4}$$

and should be prepared in a thin cylinder of Fresnel number

$$F = A/\lambda L \approx 1 \quad . \tag{6.5}$$

Note that these conditions are just the conditions derived by BONIFACIO and LUGIATO [6.1] for their pure superfluorescence. The physical significance of those conditions is not confined to the mean-field model in which they were first formulated.

In the next section, ways of achieving these conditions will be discussed and various experiments surveyed. Particular attention will be given the first experiment to clearly exhibit SF (performed in HF) and the first experiment to satisfy quite well all of the desirable conditions for SF (performed in Cs). Later sections will discuss what happens when various of these conditions are not met, e.g., when degeneracies are present, the sample length is longer than a coherence length, or the Fresnel number is large.

6.2.2 Experimental Techniques

Preparation of the initial inversion must not leave any appreciable polarization on the SF transition. This makes coherent excitation[2] of a two-level system by a coherent π pulse very difficult since any speck of dust could result in a deviation of 10^{-4} or more from an area of π for some of the atoms. Furthermore, unless the frequency width of the exciting pulse is much greater than the absorption linewidth, the π pulse will undergo self-induced transparency reshaping and incomplete inversion will result. SKRIBANOWITZ et al. [6.6,9,10] utilized incoherent saturation of two levels of a three-level system by a broadband laser, resulting in inversion between the remaining two levels. This simple preparation method was a significant step forward and is the basic idea behind all of the present methods. GROSS et al. [6.15] have used stepwise excitation via two allowed electric dipole transitions to obtain inversions on two levels of a four-level scheme. FLUSBERG et al. [6.16] have employed two-photon absorption, stimulated Raman scattering, and a stimulated three-photon interaction to prepare inverted SF systems. KARRAS et al. [6.17] have observed visible emission (which may or may not be SF) following incoherent excitation by a pulsed electrical discharge. Since an initial inversion with no macroscopic polarization is fundamental to the SF process, it is appropriate that the preparation process itself be studied carefully. BOWDEN and SUNG [6.19] have investigated theoretically the coherence developed on the SF transition by excitation with a coherent pulse.

Ideally, the inverted transition will be nondegenerate and without competing transitions. Often the branching ratios and wavelengths of competing transitions are such that one transition has the shortest τ_R by a factor of two or three and will dominate the decay. That is, the upper state population will be transferred almost entirely to the lower state of the strongest SF transition with a branching factor much closer to unity than for spontaneous emission. The strongest SF transition still possessed additional degeneracy in most experiments to date. In molecules the degeneracy may be very high, but clustering of electric dipole moment values often makes the degeneracy unimportant. In atomic systems of low degeneracy, the dipole values may differ substantially and competition results; static magnetic fields have been applied to remove such Zeeman degeneracies. Nearly degenerate hyperfine-split or fine-structure-split transitions must also be made nondegenerate if interferences such as quantum beats [6.20] are to be removed from the SF decay.

---

2 Much of the early work was approached from the point of view of SR, i.e., excitation of a two-level system by a coherent pulse. But it was recognized that excitation by as close as possible to a π pulse would result in very interesting quantum initiation and fluctuations. See, for example [6.14].

The time inequalities are satisfied by a careful choice of transition, inversion density, and sample length. Collisional contributions to T_1 and T_2' can be reduced by lowering the pressure provided adequate inversion density is still achievable. But the radiative contributions to T_1 and T_2' are inescapable. T_2^* can be reduced by using an atomic beam and perpendicular excitation, so the motion along the laser beam and hence the Doppler shift are reduced. In those cases where an external field is applied, for example to remove degeneracies, there may be contributions to T_2 from field inhomogeneities.

Once the relaxation times are made as long as possible, one should adjust τ_R to be at least 20 to 100 times smaller than the shortest relaxation time since τ_D/τ_R is typically 20 to 100. The density and length must be adjusted to satisfy this condition and the condition $\tau_E < \tau_R$. If the inversion density is limited, this may be difficult. Also the availability and response time of sensitive detectors at the SF wavelength place further restrictions on the time scale of an experiment. In addition, one must have a means of producing the inversion density quickly. Most visible and near-IR experiments have not satisfied the conditions of (6.3,4) because typical T_2^* dephasing times range from 0.1 to 5 ns and typical dye laser pulse durations are a few nanoseconds. The MIT group has suggested that high gain can reduce dephasing so that $\tau_D < \alpha L T_2^*$ is sufficient.

The Fresnel number of the SF transition is made close to unity by adjusting the diameter of the excitation beam once the length is fixed by the considerations above. Since the excitation wavelength is usually shorter than the SF wavelength the Fresnel number of the excitation is greater than one. Only recently has the desire to make quantitative comparisons of data and theory stimulated careful measurements of the Fresnel number of the inverted cylinder [6.11].

6.2.3 Survey of Experiments

We exclude from the outset the multitude of experiments in which a strongly excited system emitted a variety of coherent multicolored radiations. No doubt many of these emissions were SF, but it is almost certain that none of them satisfied closely the desirable conditions for SF. And since the relevant parameters for SF were not measured, those experiments are of little value in our present goal of carefully comparing experimental data with SF theories. The principal contributions of the relatively few remaining experiments will be briefly summarized.

The first clear observation of SF was reported in 1973 by SKRIBANOWITZ et al. [6.9]. They used a rather incoherent laser at 2.5 μm to invert an 84-μm rotational transition ($v = 1$, $J = 3$ to 2) in HF (Fig.6.1). They saturated two levels of a three-level system to achieve complete inversion between the excited level and a lower third level. The importance of this excitation method has already been stressed. They observed the important signatures of SF (Fig.6.2): coherent end-fire emission in a burst of radiation about 200 ns long delayed only 1 μs from the 100 ns

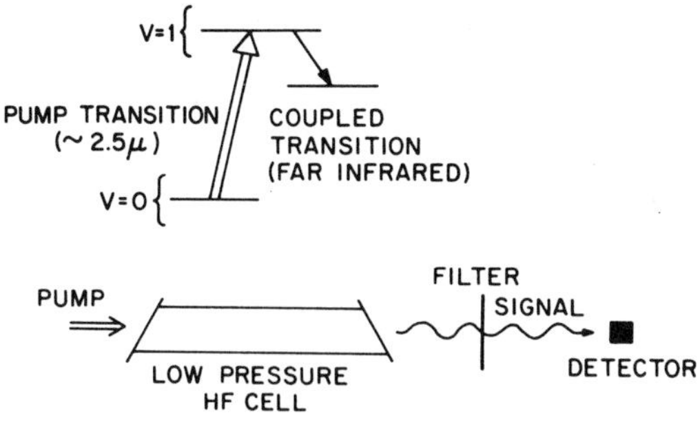

Fig.6.1. HF level scheme and schematic of experimental setup [6.10]

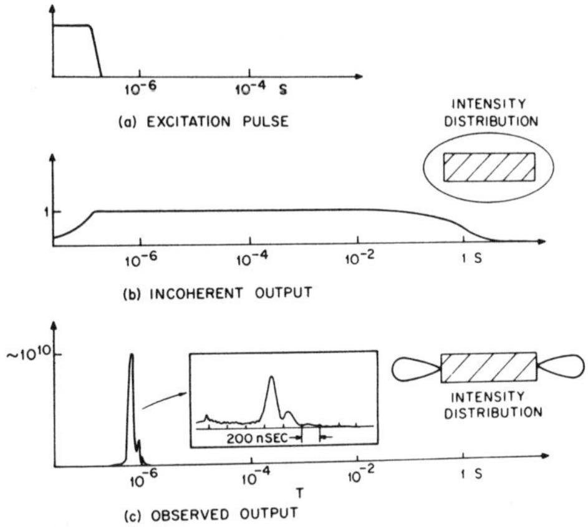

Fig.6.2a-c. Comparison of observed output and incoherent spontaneous emission in HF. Time is plotted on a logarithmic scale. (a) Pump laser pulse. (b) Output expected from incoherent spontaneous emission, exhibiting exponential decay and an isotropic radiation pattern. (c) Observed output, exhibiting ringing, a highly directional radiation pattern, and a peak intensity of $\sim 10^{10}$ times that of (b). The inset shows the time evolution of the same pulse with a linear time scale [6.10]

excitation pulse, with peak intensity proportional to N^2, the whole process being complete in a time 10^6 shorter than the ordinary spontaneous emission lifetime!

In the analysis of their experiment, the MIT group discovered the importance of propagation effects in the evolution of SF under usual laboratory conditions. They observed asymmetric outputs with long tails containing three or four modulations or rings in contrast to the symmetric pulses

$$I \propto \mathrm{sech}^2[(t - \tau_D)/2\tau_R] \tag{6.6}$$

predicted by both semiclassical and quantized field theories neglecting propagation effects. To include the fact that the field can be different in different parts of an extended sample, they applied coupled Maxwell-Bloch equations (already successful in treating propagation effects of self-induced transparency [6.21] including BURNHAM-CHIAO ringing [6.22]) to SF. Of course, these equations do not evolve from a purely inverted initial condition, so an initial tipping angle θ_0 of the polarization source is essential. The magnitude of θ_0 to be taken for SF has been under much discussion [6.4,7,8,12,23]; only recently has there been an experimental determination [6.12] which is also consistent with recent quantized treatments including propagation effects (Sect.6.4). The MIT group found, for reasonable values of θ_0

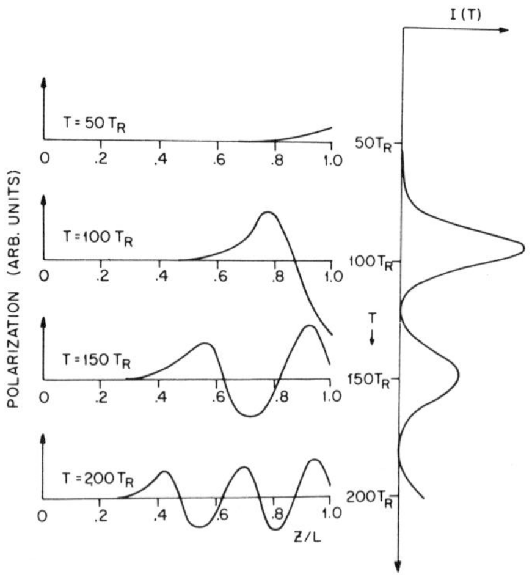

Fig.6.3. Sketch of the build-up of polarization in the HF medium, showing the polarization as a function of z at times T = 50 T_R, 100 T_R, 150 T_R, and 200 T_R. The corresponding output intensity patterns is shown at the right [6.10]. T_R is the same as the τ_R of this paper

in uniform-plane-wave simulations of SF, that the polarization is highly nonuniform spatially and the output pulse has a number of Burnham-Chiao rings (Fig.6.3). The simulations yielded ringing much stronger than observed. A linear loss term, $-K\varepsilon$ was added to the usual field equation to account for linear diffraction losses:

$$KL = \frac{1}{2}\ln(1 + F^{-2}) \quad . \tag{6.7}$$

KL = 2.5 (F = 0.08) was found to suppress the ringing to the observed level, although F closer to one was thought to be the experimental situation (Sect.6.3).

The pioneering experiment in HF by FELD and coworkers opened up experimental SF and led to a semiclassical treatment including propagation effects. Their three-level excitation scheme and realization of the importance of propagation effects are outstanding contributions. Since they have not given a detailed discussion of the accuracy with which various relevant quantities were measured, their comparisons between data and theories remain semiquantitative. ROSENBERGER et al. [6.24] have observed far-infrared SF on 496-μm rotational transitions in methyl fluoride (CH_3F) pumped by 9.55-μm pulses from a single-mode CO_2 TEA laser. Their results complement the HF experiments in that the principal relaxation mechanism was homogeneous rotational relaxation by collisions rather than the inhomogeneous Doppler broadening of the HF case. The values of T_1 and T_2' were typically 67 ns, shorter than both the 100-ns excitation pulse length and the 100- to 200-ns typical delays. Single-pulse emission with no ringing was observed, equally directed in the forward and backward directions. With such short relaxation times and small Fresnel number (0.23), ringing would not be expected. They do observe that most of the inversion energy is emitted coherently and with the characteristic N^2 dependence. They do not find good quantitative agreement between their data and the REHLER-EBERLY [6.14], BONIFACIO-LUGIATO [6.1], or uniform-plane-wave Maxwell-Bloch [6.6,9,10] approaches. Perhaps even more important than the homogeneous relaxation study, is their preliminary investigation with EHRLICH et al. [6.25] of swept-gain excitation (Sect. 6.5.1). Swept-gain coherent emission is particularly interesting as a potential source of coherent X rays [6.27].

Three groups began independently and concurrently to study SF in alkali vapors using pulsed dye lasers. GROSS et al. [6.15] used two 10-kW, 2-ns-long, 10-GHz-wide pulses from two dye lasers pumped by the same 1-MW N_2 laser to excite the 5S state in Na by two successive allowed transitions from the 3S ground state. They observed several SF emissions; particularly interesting are cascading SF on the 5S-4P, 3.41-μm and 4P-4S, 2.2-μm transitions. The expected N^2 dependence goes over to an N dependence at high densities for which quasi-stationary conditions obtain. That is, the upper level is continuously replenished as the SF pulse evolves, much the same as in many mirrorless lasers. Their experiment was the first observation of near-infrared SF. Their pulse length was not much shorter than τ_D or T_2^* and possible hyperfine quantum beats were not considered, so the reported Na data have limited value for detailed data-theory comparisons. A later experiment demonstrating Doppler beats [6.28] in SF is fascinating and is described in Sect.6.6.

FLUSBERG et al. [6.15] have observed SF on a multitude of transitions in Na, Rb, Cs and Tl. Their work is especially interesting from the variety of higher order processes employed to generate an inversion: two-photon absorption, electric quadrupole absorption, and a stimulated three-photon process consisting of stimulated Raman scattering followed by electric quadrupole absorption. They also observed delays as long as 12 T_2^*, essentially the same emission in the forward and backward directions, and complicated pulse modulation, ringing, beats, or whatever. The emission at 8184 Å brought observed SF almost into the visible.

The third concurrent alkali experiment by the present authors, which was designed specifically to make detailed comparisons between theory and experiment, will be discussed in detail in the next section.

Recently CRUBELLIER et al. [6.29] have performed an atomic beam SF experiment in Rb. They used the atomic beam to lengthen T_2^* just as in the Cs experiment [6,11,20]. They emphasized that the branching ratios between competing transitions are often very different in SF from their values in spontaneous emission, as was also noted in the Cs work. They made interesting measurements of polarization properties of the SF and gave an example of spontaneously broken symmetry.

OKADA et al. [6.30] have studied the developement of SF in a three-level system coherently excited by two-photon absorption. An initial coherence between $2^2S_{1/2}$ and $3^2S_{1/2}$ levels of lithium was prepared by two-photon excitation with 30-ps, 734.9-nm pulses. The cascade emissions via the $^2P_{1/2,3/2}$ intermediate states are at 812.6 and 670.8 nm. They investigated the evolution of the SF as a function of the atomic number density using a picosecond camera. They observed a noisy emission of discontinuous small pulses at low densities and a single large pulse (with some "rings") at high enough densities. They solved coupled Maxwell-Schroedinger equations assuming delta-function swept excitation, linear approximation in the emission stage, and a Lorentzian line broadening for Doppler effect. At relatively low densities, cooperative emission can grow from spontaneous emission, but its growth is soon suppressed by the Doppler dephasing before the uppermost level is depopulated. A succession of small pulses is then expected over a time much longer than T_2^*. This phenomenon is essentially amplified spontaneous emission (ASE). The transition from SF to ASE by increasing dephasing has recently been studied theoretically by SCHUURMANS and POLDER [6.31]. At high enough densities the emission occurs so fast by SF that the uppermost level is depleted before dephasing is effective. Their analysis then satisfactorily explains the observed behavior. It also showed that when the excitation is weak and when the two-photon excited superposition state is fully coherent, the cascade emission in the forward direction cannot grow since the polarizations of the upper and lower transitions couple destructively. The observed forward emission at 670.8 nm (lower transition) appeared only at high densities in agreement with their analysis.

MAREK [6.32] has studied beats in the SF of rubidium. Particularly noteworthy is his observation of beats from different isotopes (Sect.6.6.2).

Recently the wavelength range of well-documented SF has been pushed into the visible. BRECHIGNAC and CAHUZAC [6.23] have reported SF in europium at 557.7 nm and 545.3 nm, corresponding to the transitions 5d 6p $^{10}D_{9/2} \to$ 6s 5d $^{10}D_{11/2}$, 6s 5d $^{10}D_{7/2}$. Here excitation took place in three steps. The absorption of a first photon was followed by SF decay to an intermediate level and the absorption of a second photon. CAHUZAC et al. [6.34] have observed SF on many visible lines in europium. In particular they have investigated SF at 605.7 nm from the transition z $^8F_{9/2} \to$ a $^8D^0_{9/2}$. The excitation also involved a three-step process, two-photon absorption to $5d^2$ $^8G_{9/2,11/2}$ and subsequent SF decay to z $^8F_{9/2}$ with emission in the infrared. The possibility of using sensitive photomultipliers and photographic techniques to study SF is fascinating indeed.

6.2.4 Details of the Cesium Experiment

The experiments in Cs by VREHEN and GIBBS [6.11,20,35-40] were specifically designed to approach as closely as possible the desirable conditions for SF summarized by (6.3-5) as motivated by the discussion by BONIFACIO and LUGIATO [6.1]. In this section the satisfaction of (6.3-5) in Cs is discussed, and the resultant data are presented and compared with theories in Sect.6.3. Much of the remainder of the paper discusses the relaxation, one at a time, of several of the conditions (6.3-5).

The pertinent energy levels of Cs and the experimental arrangement are presented in Fig.6.4. SF was observed at 2.9 μm on the $7^2P_{3/2}$ to $7^2S_{1/2}$ transition after 2-ns excitation of a $6^2S_{1/2}$ to $7^2P_{3/2}$ transition. This system was selected following a survey of laser-absorber systems potentially useful in achieving near-ideal conditions for SF. By using an atomic beam to lengthen the Dopper dephasing time and

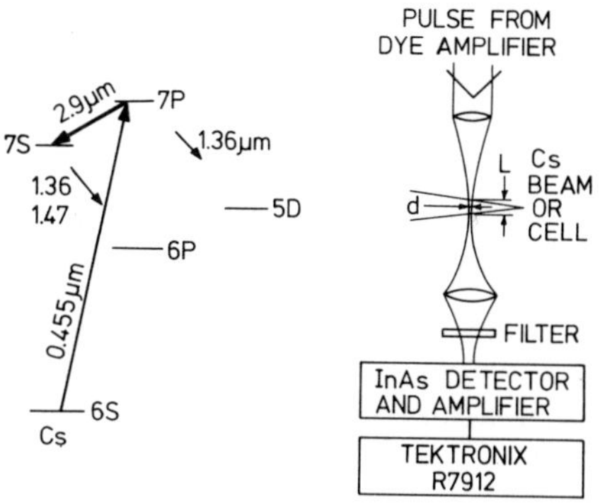

Fig.6.4. Simplified level scheme of Cs and diagram of the experimental apparatus [6.11]

a magnetic field of 2.8 kOe to remove level degeneracies, all of the desirable conditions for SF were approached closely and the important parameters were known or measured. The reader interested primarily in the data may proceed directly to Sect.6.3.

Complete inversion was obtained via the MIT scheme of saturating one transition of a three-level system. The dye laser 455-nm $6S_{1/2}$ to $7P_{3/2}$ excitation pulse of 2-ns duration and 500-MHz bandwidth was weakly focussed, yielding a peak intensity on axis of about 10 kW/cm^2. The transverse profile was always smooth and nearly Gaussian. The evidence that the saturation was complete was that the average delay time τ_D changed very little for a factor of 3 reduction in excitation intensity.

A nondegenerate two-level SF transition was prepared by selecting a particular magnetic substate in a strong magnetic field [6.36]. Without the magnetic field, several hyperfine states were simultaneously excited, giving rise to quantum beats [6.20,35] in SF (Sect.6.6). In a 2.8-kOe magnetic field, the $7P_{3/2}$ state is in the high-field Paschen-Back regime, so the good quantum numbers are the electronic and nuclear magnetic projections m_J and m_I, respectively. In that field, the transition from $6^2S_{1/2}$ ($m_J = -1/2$, $m_I = -5/2$) to $7^2P_{3/2}$ ($m_J = -3/2$, $m_I = -5/2$) was saturated without exciting neighboring transitions [6.36]. Since that upper state can only decay to $7^2S_{1/2}$ ($m_J = -1/2$, $m_I = -5/2$), a nondegenerate two-level system resulted. Competition from 7P to 6S and 5D transitions was negligible [6.35].

The sample length and inversion density were selected to satisfy (6.3). In particular, excitation of a cylinder 2 cm long perpendicular to the motion of the atomic beam yielded an escape time, $\tau_E = L/c$, of 0.067 ns and a T_2^* of 32 ns from residual Doppler dephasing and magnetic field inhomogeneities. The radiation decay times and branching ratios lead to $T_1 = 70$ ns and $T_2' = 80$ ns [6.35]. Then for a measured delay of 10 ns, τ_R of 0.5 ns was calculated from (6.1) using the measured density n of 5.5×10^{10} cm^{-3} and partial lifetime τ_0 of 551 ns calculated from the lifetimes and branching ratios [6.35]. Inequalities (6.3) were then well satisfied (all times in ns):

$$\tau_E = 0.067 < \tau_c = 0.18 < \tau_R = 0.5 < \tau_D \approx 10 < T_1 = 70 \ , \quad T_2' = 80, \ T_2^* = 32 \quad .$$

The inversion by the saturation pulse of 2-ns duration was accomplished in a short time compared to the SF build-up time τ_D of 10 ns. Of course, by controlling the beam density, τ_D was continuously varied from over 40 ns (for which it was barely distinguishable from noise) to essentially zero (for which emissions at several other wavelengths also occurred). Two other points concerning the excitation deserve consideration. First, with its spectral width about three times the transform limited value, the pump pulse is neither coherent nor completely incoherent. Its peak power fluctuates from shot to shot. Even if pulses of constant peak power are selected the corresponding SF delay times show fluctuations which exceed the expected quantum fluctuations (Sect.6.4). The average delay time, however, decreases monotonically with increasing pump power and converges to an asymptotic value. No changes in this behavior were observed when the pump pulse duration was increased to 3 ns and the spectral width to 1200 MHz. We believe therefore that the sample preparation corresponds more closely to saturation than to coherent excitation. Second, the excitation was "swept" in the sense that the SF cylinder was excited by a pulse travelling along the axis. However, the spatial length of the exciting pulse was 60 cm FWHM, much longer than the 2-cm cylinder length. In that sense, the excitation was not swept since all atoms were subjected to approximately equal fields throughout the preparation process. That the excitation impressed no preferential direction upon the inverted cylinder was verified by the failure to detect any difference between co- and counter-propagating SF emissions (within statistical and quantum fluctuations, Sect.6.4.3, [6.37]). Of course, at very high densities the emission begins during and may be preferentially along the excitation.

The final desirable condition for SF was achieved by adjusting the focussing of the excitation beam to produce an inverted cylinder of Fresnel number one *for the SF wavelength.* Since the pump wavelength was shorter than the SF wavelength, this could be done easily in principle; but determination of the diameter of an inverted cylinder prepared by a saturating pulse is nontrivial. From the possible variations in pump intensity it was estimated that the actual Fresnel number was between 0.5 and 2.0.

With the desirable condition for SF satisfied, one is in a position to observe the SF process under ideal conditions. Nevertheless, to compare the data meaningfully with theories, one must know the relevant parameters. The basic parameter is τ_R which contains information about the strength of the SF transition (through τ_0), the diffraction properties of the SF emission (through λ and L), and the density of initially inverted radiators n. The density n was the difficult quantity to measure in the Cs experiment. The total ground-state density was determined from the known geometry, initial quantity of Cs, and calculated channeling factor along with measurements of the oven temperature [6.36]. This method was checked by inserting a copper collecting plate in the beam and determining the condensed mass by several techniques of analytical chemistry [6.36]. Estimated errors in n assuming complete saturation of the pump transition were quoted as (-40, +60)%.

6.3 Single Pulses

6.3.1 Observation of Single Pulses in Cesium

Under the ideal conditions described in Sect.6.2.4, single-pulse emission was observed in Cs without a trace of ringing [6.11]. An example of one of the most nearly symmetric pulses is compared with a sech^2 in Fig.6.5. More typical output pulse shapes, shown in Fig.6.6, are asymmetric with a slower fall than rise time. But the symmetric pulses are often seen and are narrower than the asymmetric pulses emitted at the same density. Therefore, it is unlikely that the asymmetric pulses are averaged ringing pulses. The multiple-pulse outputs at the top of Fig.6.6 do not arise from ringing (Sect.6.3.4).

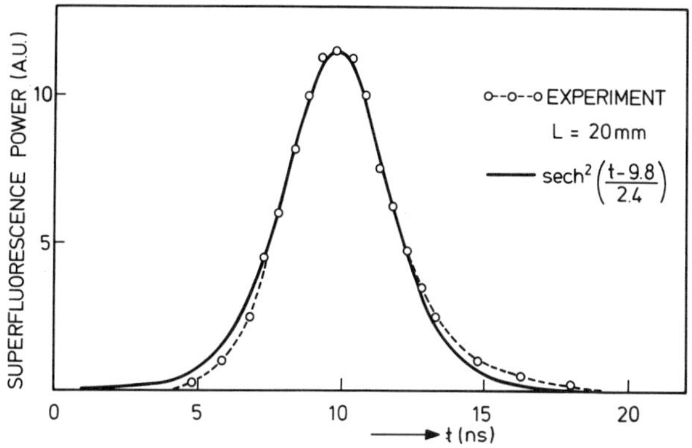

Fig.6.5. Example of the very symmetrical pulses that have often been observed in Cs [6.36]

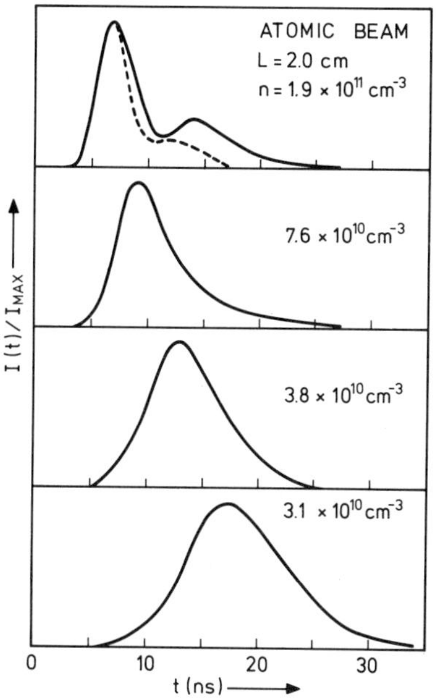

Fig.6.6. Normalized single-shot pulse shapes for several Cs densities n; Fresnel number $F \approx 1$. Uncertainties in the values of n are estimated to be (+60, -40)% [6.11]

The lack of ringing, which conflicts with uniform-plane-wave simulations discussed in Sect.6.3.2, is difficult to explain in Cs. This is because the usual scapegoats are absent; all the relaxation times T_1, T_2', and T_2^* are too long; the Fresnel number is known to be close to unity; the transition is nondegenerate, etc. In the other SF experiments, a regime of single-pulse emission usually occurred, but relaxation and diffraction may have prevented ringing.

6.3.2 Maxwell-Bloch Equations and Ringing

In Sect.6.2.3 it was noted that the MIT group introduced propagation effects into the theory of SF [6.6,9,10], i.e., they did not require that the polarization and field have the same values at every position in the superfluorescing sample. They accomplished this by utilizing the semi-classical treatment already successfully applied to a multitude of propagation problems [6.21,22,41]. Namely, the matter is quantized and described by Schroedinger's equation which, in Bloch's notation, becomes

$$\dot{u} = +(\omega_0 - \omega)v - u/T_2' \quad , \tag{6.8}$$

$$\dot{v} = -(\omega_0 - \omega)u - v/T_2' - w\kappa\varepsilon \quad , \tag{6.9}$$

$$\dot{w} = -(w + 1)/T_1 + v\kappa\varepsilon \quad , \tag{6.10}$$

and the electromagnetic field E is not quantized, but assumed to be a classical coherent wave

$$E(z,t) = \varepsilon(z,t)\, e^{i[\omega t - kz - \phi(z)]} + \text{c.c.} \tag{6.11}$$

obeying Maxwell's travelling-wave equation

$$\frac{\partial\varepsilon}{\partial z} + \frac{1}{c}\frac{\partial\varepsilon}{\partial t} = -\frac{2\pi\omega}{c}\, npv \quad . \tag{6.12}$$

Slowly varying envelope, rotating wave, uniform-plane-wave, and forward-only approximations have been made to arrive at these equations. The polarization has in-phase dispersive component u and out-of-phase absorptive component v:

$$P(z,t) = [u(\Delta\omega,z,t) - iv(\Delta\omega,z,t)]e^{i[\omega t - kz - \phi(z)]} \quad , \tag{6.13}$$

$$u = (\rho_{ab} + \rho_{ba})/2 \quad , \tag{6.14}$$

$$v = i(\rho_{ba} - \rho_{ab})/2 \quad , \tag{6.15}$$

$$w = \rho_{aa} - \rho_{bb} \quad , \tag{6.16}$$

$$\kappa = 2p/\hbar \quad . \tag{6.17}$$

Here, ρ is the density matrix of the two-level SF transition with upper state a and lower state b. The dipole moment p of the SF transition is related to τ_0 by [6.35]

$$\frac{1}{\tau_0} = \frac{8}{3}\frac{\omega^3 p^2}{\hbar c^3} \quad . \tag{6.18}$$

A semiclassical description is inherently incapable of providing information on the quantum statistics of the emitted SF pulses (Sect.6.4.3). Even more serious for a comparison with experimental pulse shapes is its failure to evolve at all from a purely inverted state. For then $u(t = 0) = v(0) = 0$ and $\varepsilon(z,0) = 0$, so by (6.8-12) $\dot{u} = \dot{v} = \dot{w} = \varepsilon = 0$ for all time. The actual initiation of SF is, of course, a quantum process since it begins as ordinary incoherent spontaneous emission with the usual angular distribution determined by the angular momenta of the initial state. Traditionally the entire quantum initiation process is collapsed into an effective initial tipping angle θ_0 or a random polarization source throughout the sample [6.6,9,10].

The concept of tipping angle is readily grasped by solving (6.7,8,11) on resonance ($\omega = \omega_0$) and with no relaxation ($T_1 = T_2' = \infty$); then $u = 0$,

$$v = -\sin\theta \tag{6.19}$$

$$w = \cos\theta \tag{6.20}$$

where $w(0) = +1$ and

$$\theta = \kappa \int_{-\infty}^{t} \varepsilon dt \tag{6.21}$$

is the electric field area and is the angle through which the polarization vector (u,v,w) is rotated in the v,w plane by the on-resonance light field. Therefore, in a retarded frame $T = (t - z/c)/\tau_R$ and $Z = z/L$

$$\frac{\partial\theta}{\partial T} = \kappa\tau_R\varepsilon \quad , \tag{6.22}$$

$$\frac{\partial^2\theta}{\partial Z\partial T} = \kappa\tau_R \frac{\partial\varepsilon}{\partial Z} = +\sin\theta \tag{6.23}$$

using (6.1,12,18). Equation (6.23) is the sine-Gordon equation. If spatial variations are negligible, then (6.23) can be solved to yield an intensity proportional to N^2 with the sech^2 time dependence of (6.6). But spatial variations are important under the conditions of all SF experiments to date, so the numerical solutions of (6.23) by BURNHAM and CHIAO [6.22], McCALL [6.42], MacGILLIVRAY and FELD [6.6,9,10], and BULLOUGH et al. [6.43] are especially useful.

6.3.3 Comparison with Computer Simulations

For $10^{-8} < \theta_0 < 10^{-3}$, the range of experimental values calculated from the now-accepted value (Sect.6.3)

$$\theta_0 = 2/\sqrt{N} \quad , \tag{6.24}$$

numerical solutions of the sine-Gordon equation or, equivalently, the coupled Maxwell-Bloch equations result in much stronger ringing than observed. For example, in the Cs experiment for which the parameters needed for the simulations are relatively well known, too much ringing is predicted (Fig.6.7).

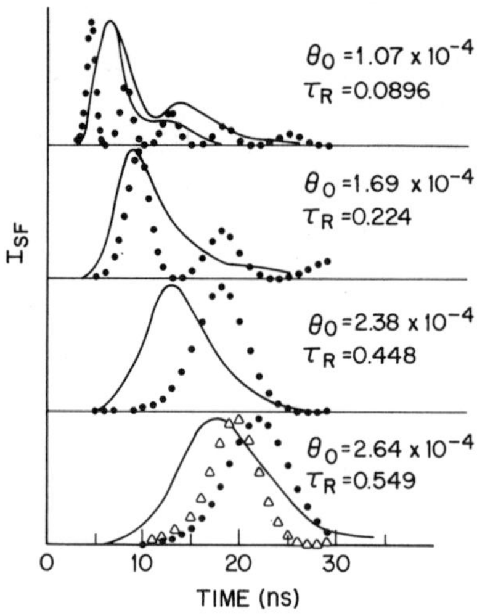

Fig.6.7

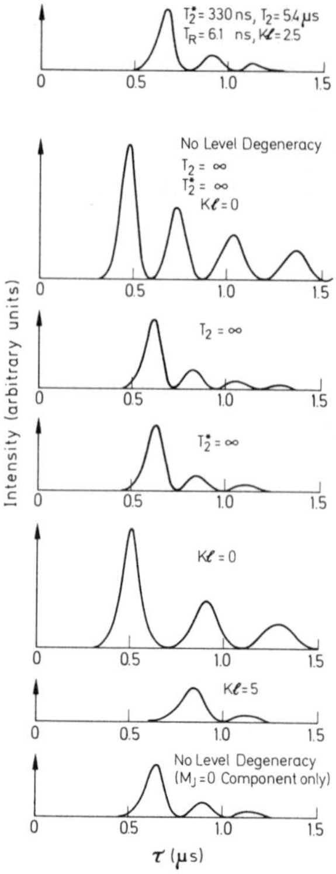

Fig.6.8

Fig.6.7. Comparison of normalized Cs data (solid curves) with simulations of one-way Maxwell-Bloch equations initiated by a short input pulse of area θ_0. The experimental values of $\tau_R = 8\pi\tau_0/3n\lambda^2L$ are from the top down, 0.14, 0.35, 0.71, and 0.87 ns; the τ_R values used in the fits are about 63% of the experimental values as required so that θ_0 could be given by $2/\sqrt{N}$ and some agreement with observed delays still obtained. Experimental relaxation times are used except for the triangular point curve with 10^5 ns [6.23,38]. On the lowest two curves the ringing is not shown, but it is still sizable

Fig.6.8. Computer results showing the influence of parameters on pulse evolution. The uppermost curve is a theoretical fit resembling the HF data closely. All parameters have the same values as in this curve except when stated otherwise. The values of the modified parameters are indicated in the figure. The same intensity scale is used throughout [6.10]. Clearly, choosing KL = 2.5 is the most important step in reducing the strong ringing to the observed level

Various possible effects for reducing the simulated ringing to the observed level have been investigated. The MIT group [6.6,9,10] found that HF degeneracy and relaxation effects were relatively small (Fig.6.8). But by introducing a linear loss term $-K\varepsilon$ to the right side of (6.12) to account for linear diffraction losses, they found that $KL \approx 2.5$ suppressed the ringing to the observed level. Their formula (6.7) for KL in terms of the Fresnel number F yields F = 0.08 for KL = 2.5. They were not explicit enough about their experimental F to make possible a quantitative comparison, but an F of about unity was mentioned. In the Cs case, F is known to be about one, so that a large KL cannot be justified to suppress ringing.

The use of a large θ_0, of the order 10^{-2}, does suppress the ringing and brings the observed Cs delays into better agreement with the simulations [6.38]. However, after the recent experimental and theoretical work described in Sect.6.4, the value $\theta_0 \approx 2/\sqrt{N}$ is now accepted, ruling out large θ_0's for Cs. Instead, $\theta_0 \approx 10^{-4}$ and large ringing results (Fig.6.7).

Two-way simulations have been made to evaluate the effect of two-way competition upon the amount of ringing [6.6,9,10,43]. The smaller θ_0, the less effect there is upon the ringing because the first few rings out of each end are emitted by a polarization primarily situated near the emitting end, i.e., at first the polarization is largest near the ends and is directed outward on both ends, so little interaction between the two oppositely directed emitting regions occurs. Even for $\theta_0 = 0.032$ (their $\delta = 10^{-3}$), SAUNDERS and BULLOUGH [6.43] found only a 30% decrease in the height of the first ring when two-way effects are included (Fig.6.9). And, of course, as just discussed, θ_0 is now believed to be much smaller, $\approx 10^{-4}$, so two-way effects should be even smaller.

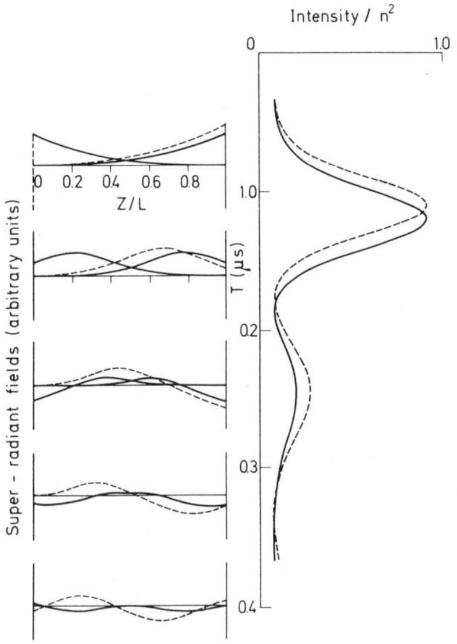

Fig.6.9. Comparison of uniform-plane-wave simulations under one-way and two-way conditions, $\delta = \theta_0^2$ [6.43]

At this writing, there is still no simulation satisfactorily in agreement with the Cs data. The pulse delays are in fair agreement within experimental uncertainties in density. Even the pulse widths are not very much narrower than observed. But the predicted ringing is clearly too large. The present likely explanation is that simulations are made neglecting transverse variations in the input and in the evolution: an incident uniform plane wave (UPW) is assumed and no radial dependence is included in the equations of motion. Ideal SF initial conditions are always nonuniform transversely; a Fresnel number F of one is needed to prevent the poor lengthwise communication of a small F cylinder and the mode competition of a large

F cylinder. Gaussian averaging of UPW solutions does not give good agreement either: ringing is largely eliminated, but the remaining asymmetry far exceeds that observed. Simulations with dynamic transverse effects included require phase and radial evolution in the equations of motion. They have not yet been made for SF. However, dynamic transverse effects are known to dominate the evolution of pulses undergoing self-induced transparency under similar conditions [6.44].

The earlier theories, including the Bonifacio-Lugiato quantized mean-field theory [6.1], predict single-pulse emission, but the predicted widths for Cs (Fig.6.5) are a factor of two narrower than observed. Furthermore, the simulations of coupled Maxwell-Bloch equations clearly indicate that the actual field within the sample is far from slowly varying as needed to represent it by a mean field (Fig. 6.3) [6.6,9,10,43]. With the recent success in bridging the gap from a fully quantum and linear initiation regime to a semiclassical and later nonlinear regime, one is able to obtain information on fluctuations from quantum statistics while retaining the propagational features of the coupled Maxwell-Bloch approach [6.4,7,12]. It remains to be seen if simulations with transverse effects included will reproduce the data, thus explaining the discrepancy between plane-wave simulations and observations.

6.3.4 Multiple-Pulse SF and Transverse Effects

For the highest Cs densities multiple-pulse output was observed as shown by the uppermost curve in Fig.6.6 [6.11]. The amount of "ringing" varied considerably from shot to shot, far more than the lower-density pulses fluctuated.

It is now believed that this multiple pulsing is not BURNHAM-CHIAO ringing [6.6,9,10,22], which arises from uniform-plane-wave simulations (presumably such ringing is obscured by Gaussian averaging and dynamic transverse effects as discussed in Sect.6.3.3). Nor is it the oscillatory SF predicted by BONIFACIO and LUGIATO [6.1], which should become more and more pronounced at higher densities.

Multiple pulsing in Cs arises from transverse effects [6.37]. The observed pulse shapes depended on the position of the detector in the image plane (Fig.6.10). With the help of a beam splitter, two images were formed; a detector (diameter 150 μm) was placed in each image and the signals of both detectors were displayed on the same trace of the Transient Digitizer (one signal was suitably delayed). When the detectors occupied equivalent positions in the image planes, their signals were very similar. When the detectors occupied nonequivalent positions, they showed quite different signals. Examples are given in Fig.6.10. The experiment suggests that the multiple pulses cannot be described in a plane-wave approximation. The shapes of the single pulses observed at longer delays do not depend on detector position significantly. The tendency toward multiple-pulse generation increases with length L for constant $F \approx 1$ [6.36], but it decreases with increasing F at constant L (Sect.6.5).

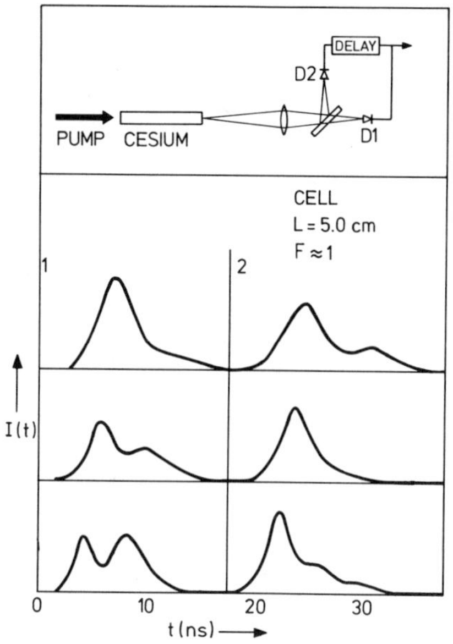

Fig.6.10. Pulse shapes in Cs observed simultaneously with two detectors at different positions in the image plane, as indicated at the top. Cell, L = 5.0 cm, F ≈ 1 [6.37]

The onset of multiple pulsing is roughly for $\tau_E \geqq \tau_R$, i.e., when the sample length L exceeds the cooperation length, L_c, multiple pulsing sets in. A possible explanation is that SF occurs in a portion of the sample length before the entire sample can communicate and cooperate. If the effective emitting length is less than L and F = 1 for the entire sample, then F > 1 for the emitting portion. This implies that more than one mode can superfluoresce. The lack of pulse-to-pulse reproducibility could then be explained as fluctuations arising from competition between modes and variations introduced by fluctuations in the quantum initiation process. For much higher Fresnel numbers, the number of competing modes and number of competing segments become so large that every shot is already an average with small fluctuations and little evidence of multiple pulsing.

Ringing was seen in the MIT [6.6,9,10] (Fig.6.2c) and Paris [6.15] experiments also for a length about equal to L_c. It seems likely that their ringing was not BURNHAM-CHIAO ringing either, but multiple pulsing from transverse effects. MacGILLIVRAY and FELD [6.6] suggest that $\tau_E < \tau_D$ should be the appropriate condition (rather than $\tau_E < \tau_R$) to avoid cooperation length problems. Since $\tau_D \approx 20\tau_R$ in the Cs experiment, clearly multiple pulsing occurred at a density an order of magnitude lower than their criterion would suggest.

6.4 The Initiation of Superfluorescence

6.4.1 Theory

Superfluorescence is initiated by quantum fluctuations, at least for frequencies in the visible and near-infrared where thermal radiation can be neglected. Fully retarded quantum-mechanical descriptions of the initiation of SF have been provided recently by GLAUBER and HAAKE (GH) [6.7] and by POLDER, SCHUURMANS and VREHEN (PSV) [6.4,8]. In the GH treatment normally ordered correlation functions are calculated and in those correlation functions the quantum noise appears as the zero-point fluctuations of the matter. In the work of PSV antinormally ordered correlation functions are calculated and the quantum noise then originates in the zero-point fluctuations of the field. The two approaches are equivalent and complementary. They are, in fact, different specializations of a more general theory, as shown by SCHUURMANS and POLDER [6.45]. In the remainder of this section we follow their arguments. First consider the Bloch equations (6.8-10). During the initiation these equations can be linearized since the inversion is nearly constant, $w = 1$ and $\dot{w} = 0$.

For on-resonance field only, $\omega = \omega_0$, and in the complete absence of relaxation, $T_1 = T_2' = \infty$, (6.8,9) can be combined into

$$\dot{u} - i\dot{v} = i\kappa\varepsilon \tag{6.25}$$

where $(u - iv)$ is a properly normalized complex polarization amplitude and ε is now a complex field amplitude. With the substitutions $P = (u - iv)$, $E = i\kappa\tau_R\varepsilon$, $T = (t - z/c)/\tau_R$, and $Z = z/L$ the Maxwell-Bloch equations read

$$\frac{\partial P}{\partial T} = E \quad , \tag{6.26}$$

$$\frac{\partial E}{\partial Z} = P \quad . \tag{6.27}$$

These same equations can be derived quantum mechanically [6.4]. E and P are then Heisenberg operators on the initial state of the systems, i.e., the state $|\psi\rangle$ with all atoms excited and no photons present. These equations (6.26,27) are completed with an initial condition ($T = 0$) for P and a boundary condition ($Z = 0$) for E. Since the atomic system is completely inverted at $T = 0$ one finds for the initial operator polarization $P(Z,T = 0) = P_0(Z)$ that

$$P_0^+|\psi\rangle = 0 \tag{6.28}$$

and furthermore the commutator

$$[P_0^+(Z),P_0(Z')] = \delta(Z - Z')/N \tag{6.29}$$

illustrating that P_0 is a Bose operator.

Only SF emitted to the right is considered and consequently the electric field at the left-end face equals the vacuum field E_0 incident on it, i.e., $E(Z = 0,T) = E_0(T)$ with the Bose operator E_0 satisfying

$$E_0|\psi\rangle = 0 \tag{6.30}$$

and furthermore

$$[E_0(T),\ E_0^+(T')] = \delta(T - T')/N \quad . \tag{6.31}$$

Equations (6.26-31) provide a fully quantum-mechanical description of the initiation of SF allowing the calculation of arbitrary correlation functions of field and matter. To obtain vanishing initial and boundary variables SCHUURMANS and POLDER [6.45] introduced the matter field $\Omega = E - E_0(T)$ and the collective Bloch vector component $M = P - P_0(Z)$. The Maxwell-Bloch equations (6.26-27) then read

$$\frac{\partial M}{\partial T} = \Omega + E_0(T) \quad , \tag{6.32}$$

$$\frac{\partial \Omega}{\partial Z} = M + P_0(Z) \quad . \tag{6.33}$$

In these equations the operators $E_0(T)$ and $P_0(Z)$ act as sources on which M and Ω depend linearly. Since $P_0^+|\psi\rangle = E_0|\psi\rangle = 0$ it follows that $P_0(Z)$ does not contribute to any antinormally ordered correlation functions and $E_0(T)$ does not contribute to normally ordered correlation functions. For a description of the matter, concerning antinormal ordering, the mean-squared tipping angle of the collective Bloch vector equals $4\langle MM^+\rangle = 4\langle PP^+\rangle$. For this ordering, which was adopted by PSV, the quantum fluctuations appear as zero-point fluctuations of the field. For a calculation of the field, normal ordering is appropriate since the mean-field intensity is $N\hbar\omega_0\langle\Omega^+\Omega\rangle/\tau_R$. For this ordering, which was adopted by GH, the quantum fluctuations appear as the zero-point fluctuations of the matter.

SCHUURMANS and POLDER further demonstrated how two different descriptions using stochastic variables can be formulated depending on the ordering scheme. For antinormal ordering Ω, Ω^+, and M, M^+ are treated as complex valued c numbers $\tilde{\Omega}$, $\tilde{\Omega}^*$, and $\tilde{M}$, $\tilde{M}^*$. (E_0, E_0^+) is considered as a classical fluctuating field source $(\tilde{E}_0, \tilde{E}_0^*)$ and P_0 is put equal to zero. Moreover, the quantum-mechanical average and the average over stochastic variables are identified. The atom-field system is driven by a bivariate field noise source $(\tilde{E}_0, \tilde{E}_0^*)$, which is Gaussian since E_0 is a Bose operator. The second-order correlation function is

$$\langle\tilde{E}_0(T)\tilde{E}_0^*(T')\rangle = \delta(T - T')/N \quad . \tag{6.34}$$

In this stochastic variables picture the behavior of the collective Bloch vector in each individual experiment (single shot) is determined by one representative out of all possible noise source functions $E_0(T)$,

$$\tilde{M}(Z,T) = \int_0^T dT' I_0[2\sqrt{Z(T - T')}]\tilde{E}_0(T') \tag{6.35}$$

where I_0 is the modified Bessel function of zeroth order. The average over the ensemble of stochastic variables must be understood as the average over many repeated experiments. This interpretation of the individual experiment was first explicitly stated by POLDER et al. in the PSV paper [6.4]. This interpretation allows the calculation both of the average quantities and of the fluctuations. For example,

$$<\tilde{M}\tilde{M}^*>(Z,T) = \frac{1}{N}\int_0^T I_0^2(2\sqrt{ZT'})dT' \quad . \tag{6.36}$$

The effective initial tipping angle θ_0 (Sect.6.3.2) is calculated by PSV as

$$\theta_0^{PSV} = \frac{2}{\sqrt{N}}[\ln(2\pi N)^{1/8}]^{1/2} \tag{6.37}$$

and the relative standard deviation in the delay time as

$$\Delta\tau_D = 2 \cdot 3/(\ln N) \quad . \tag{6.38}$$

A similar stochastic variables description can also be given for normal ordering of the operators. The system is then driven by a bivariate Gaussian polarization noise source.

The theory sketched above gives a complete one-dimensional description of the initiation of SF in the linear regime. The theory can be extended beyond the linear regime provided the motion becomes classical in that regime, i.e., the noise becomes of minor importance for the further evolution of the system. In that case the solutions can be extended into the nonlinear regime by using classical nonlinear Maxwell-Bloch equations.

HAAKE et al. [6.46] have numerically calculated single-shot field intensities using the nonlinear Maxwell-Bloch equations with the polarization noise source of (6.33), using the parameters of the cesium experiment. Their result for the delay time fluctuations agrees with (6.38). The value $\theta_0 \approx 2/\sqrt{N}$ seems now to be accepted. However, the value to be given to θ_0 had previously been subject of much debate. BONIFACIO and LUGIATO [6.1] had arrived at essentially the same value in the framework of the mean-field theory

$$\theta_0^{BL} = \sqrt{2/N} \quad . \tag{6.39}$$

A much larger value was implicit in the work of REHLER and EBERLY [6.14]

$$\theta^{RE} = 1/\sqrt{\mu N} \tag{6.40}$$

where $\mu = 3\lambda^2/8\pi A$, and A is the cross-sectional area of the sample for Fresnel number 1. A smaller value had been predicted by MacGILLIVRAY and FELD [6.6]

$$\theta_0^{MF} = [(2\pi)^{1/2}N(\alpha L)^{3/2}]^{-1/2} \tag{6.41}$$

where αL is the steady-state amplitude gain of the sample. The θ_0 values of (6.40, 41) differ by almost four orders of magnitude in the case of the cesium experiment. The discrepancy prompted a direct measurement of θ_0 as described in the next section.

The delay time fluctuations had previously been considered by DEGIORGIO [6.47] in the mean-field approximation. He arrived at

$$\Delta\tau_D = 1.2/(\ln N) \tag{6.42}$$

which may be compared with (6.38). The difference can be attributed to the fact that the delay time depends quadratically on $\ln\theta_0$ when propagation effects are taken into account, and linearly in the mean-field theory.

6.4.2 Direct Measurement of θ_0

The direct measurement of θ_0 as reported by VREHEN and SCHUURMANS [6.12] is based on the following idea. A small area coherent pulse at the SF wavelength is injected into the SF sample immediately after it has been excited by the laser pump pulse. If the area of the injection pulse θ is smaller than θ_0 the SF emission will still be initiated by quantum noise and the average delay time will not be affected. However, if θ is larger than θ_0 the superradiant decay will be induced by the injection pulse and the average delay time will decrease. Thus by measuring the delay time as a function of θ the value of θ_0 can be determined. The experimental setup is shown in Fig.6.11. Two cesium cells are successively pumped by the same laser pulse. In the first cell, which has a high density, a SF pulse is generated with a delay of about 1.5 ns and a width of about 2 ns. This pulse serves as the injection pulse for the second cell, which has a lower density, so that without injection it emits an SF pulse with a delay of about 13 ns. The first cell roughly emits a π pulse. At the entrance of the second cell the area is smaller because of the beam divergence and because of the presence of infrared attenuators between the two cells. In the experiment the densities in both cells are kept fixed, the attenuation of the injection pulse is varied and the average delay time is measured as a function of θ. A result is shown in Fig.6.12 where the average delay time τ_D is plotted versus $[\ln(\theta/2\pi)]^2$. For $\theta<5\times10^{-4}$ the delay time is essentially independent of . For $\theta>5\times10^{-4}$ the delay time varies linearly with $[\ln(\theta/2\pi)]^2$ as might be expected from the theoretical work of BURNHAM and CHIAO [6.22]. From a number of such experiments the most probable value is found to be $\theta_0 = 5\times10^{-4}$, with an uncertainty of a factor of 5, $1\times10^{-4}<\theta_0<2.5\times10^{-3}$.

From the experimental parameters the theoretical θ_0 values calculated with (6.37,39-41) are: $\theta_0^{RE} = 2.7\times10^{-2}$, $\theta_0^{PSV} = 2.3\times10^{-4}$, $\theta_0^{BL} = 1\times10^{-4}$, and $\theta_0^{MF} = 6.7\times10^{-6}$. The measured value is in good agreement with the PSV theory, and is equally in agreement with GH, although these authors did not calculate θ_0 explicitly.

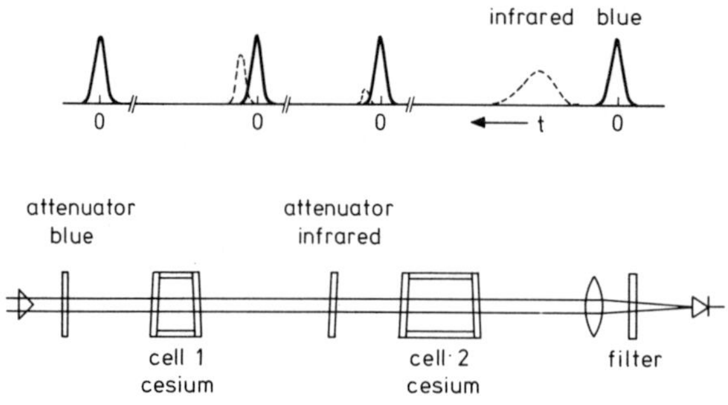

Fig.6.11. Setup for the measurement of the effective initial tipping angle [6.12]

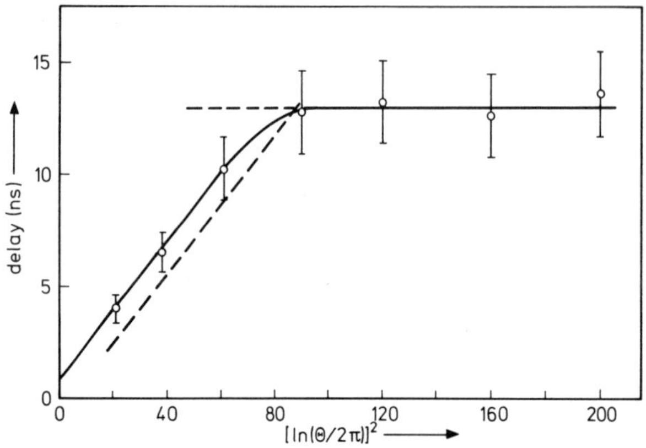

Fig.6.12. Delay time τ_D of the SF output pulse in Cs vs$[\ln(\theta/2\pi)]^2$. The dashed line is used to correct for the delay of the injection pulse with respect to the pump pulse [6.12]

An area of 5×10^{-4} for a pulse of 2-ns width corresponds to a pulse energy of 1 photon passing through the cross section of the SF sample. It may thus be said that SF is initiated by the first photon emitted along the axis of the sample.

As already noted (Sect.6.3.3) the delay times calculated with one-dimensional Maxwell-Bloch simulations using the experimental value of θ_0 are in fair agreement with the observed delay times within the experimental uncertainties in the densities. The complete absence of ringing, however, is not consistent with $\theta_0 = 5\times10^{-4}$.

The experiment provides an example of coherent pulse propagation in a narrow-band amplifier. It would be of interest to study that process with a large diameter of the pumped volume in the second cell to approach plane-wave conditions. Possibly, true BURNHAM-CHIAO ringing might then be observed, just as complete pulse break-up in self-induced transparency could only be seen with plane-wave conditions well satisfied [6.48]. Large injection areas would probably be required since it is known from experiment [6.37] (Sect.6.5.2) that SF delay times decrease with increasing Fresnel number, suggesting that the quantum noise is relatively stronger for larger F.

6.4.3 Quantum Fluctuations

The measurement of θ_0 provides information on the average strength of the quantum noise source. However, as explained in Sect.6.4.1, each single-shot experiment corresponds to one representative of all possible noise source functions. Quantum fluctuations will occur in the delay time and the pulse shape. In principle such fluctuations can be measured by simply repeating the experiment many times and analyzing the statistics. In practice the measurement is complicated by the fact that the preparation of the sample is not sufficiently reproducible because of instabilities in the pump laser pulse. Delay time fluctuations caused by these instabilities can easily dominate over the quantum fluctuations. One of the present authors [6.49] has recently overcome this difficulty by simultaneously pumping two identical cells with the same laser beam after it had been divided with a beam splitter and measuring the delay times of the two cells separately. Delay time differences at the same shot are then entirely due to quantum fluctuations. The experimental apparatus is sketched in Fig.6.13. The two pulses are recorded on the same trace of a Tektronix Transient Digitizer and stored on a memory disc. It is found experimentally that the delay times of the two cells are highly correlated. The correlation of about 80% may be attributed to fluctuations in the preparation of the sample, in particular of the upper-level density, which are common to both cells. The uncorrelated part may be ascribed to the quantum fluctuations. The analysis is complicated by the density fluctuations but it can be shown [6.49] that the standard deviation in the delay time due to quantum fluctuations can be derived from the data, nevertheless. The result of the measurements so far performed is (13 ± 3)%. This number may be compared with the theoretical estimates. For the present experiment DEGIORGIO's estimate (6.42) yields $\Delta\tau_D$ = 6.3%. The analytical expression (6.38) by POLDER et al. [6.4] gives $\Delta\tau_D$ = 12.5%, and HAAKE et al. [6.46] found numerically $\Delta\tau_D$ = 12%. The preliminary experimental results are thus in agreement with the PSV and GH theories. The experiments are continuing in order to improve the statistics.

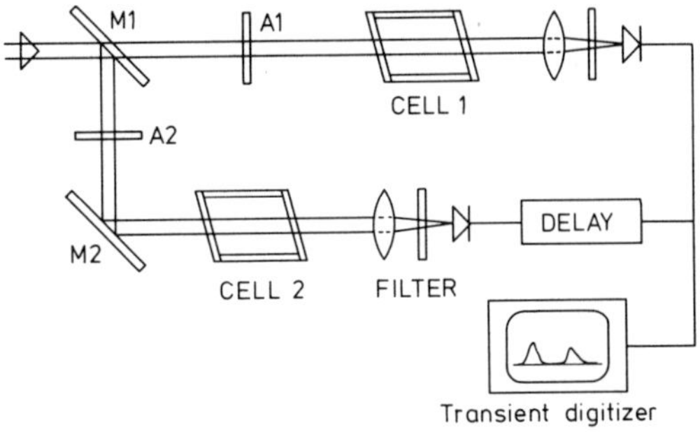

Fig.6.13. Apparatus for the measurement of quantum fluctuations in the SF delay time [6.49]

The experiment described above has been made for Fresnel number 1. It will be interesting to perform the experiment for larger Fresnel numbers. Since both the number of atoms and the solid angle in which they can radiate collectively increases linearly with increasing F, the number of photons emitted spontaneously in the first τ_R within that solid angle increases as F^2. It may be expected therefore that the quantum fluctuations in the delay time decrease with increasing F.

It is also of interest to study experimentally the delay time differences between the pulses emitted in the forward and in the backward direction. From (6.35) it is evident that the SF pulse emitted to the right (left) is initiated by the vacuum field wave travelling to the right (left) and entering the sample at the left- (right-)end face. The two vacuum field waves are statistically independent. Thus the backward and forward SF waves are uncorrelated except for a possible coupling between them through the nonlinearity of the medium. The correlation between the delay times of pulses emitted from the same sample in the forward and in the backward direction has been studied experimentally by VREHEN et al. [6.37]. The measured correlation is comparable to that between the delay times of pulses emitted in the forward direction from different, but simultaneously pumped cells. The latter is attributed to the instabilities in the pump pulse. The experiment suggests that the coupling between the forward and backward waves does not contribute significantly to the correlation between their delay times. Definite statements, however, must await further and more precise measurements.

6.5 Sample Dimensions

In the preceding two sections experiments have been reviewed that were designed to test the basic theory for the simplest possible conditions, such as nondegenerate two-level system, Fresnel number 1, L/c small compared to τ_R, negligible relaxation, etc. In later sections experiments will be discussed in which those conditions are relaxed.

The present section deals with the sample dimensions. The samples have circularly cylindrical shape. The dimensions play a part through the length L in comparison with the distance light travels in the SF characteristic time τ_R or the delay time τ_D and through the Fresnel number $F = A/\lambda L$. In an experiment L is always known with good precision. The actual value of F, however, is somewhat uncertain (Sect.6.2.4). An uncertainty of a factor 2 must be reckoned with.

6.5.1 Sample Length

If the medium is excited by a short optical pulse travelling with the velocity of light along the sample axis, then the forward SF emission can be described in terms of the retarded time $\tau = t - z/c$. One-dimensional Maxwell-Bloch theories predict that the sample length will be of no consequence for the evolution of the forward SF pulse [6.4,6,9,10] provided the interaction between forward and backward waves can be neglected. The backward wave, however, will be affected by the sample length because quantum noise waves originating from the far end will reach the front end only a time 2L/c after it was first excited. On the basis of this argument one expects a significant change in the backward pulse for $2L/c \gtrsim \tau_D$, where τ_D is the delay time of the forward pulse. Through its coupling to the backward wave the forward wave will then be modified too, but less dramatically.

The experiments on single-pulse SF in cesium revealed an interesting dependence of the pulse shapes on the sample length. For a given delay time the tendency for the formation of multiple pulses increased with the length [6.36]. Broadly speaking, single pulses were observed for $\tau_R > 2L/c$ whereas multiple pulses occurred for $\tau_R < 2L/c$. In Sect.6.3.4 evidence has been presented that the multiple pulses are related to transverse nonuniformities in the emitted intensity. They cannot be understood from a one-dimensional theory.

Recently EHRLICH et al. [6.25] have reported a strong variation of the forward-to-backward peak intensity ratio with increasing pressure in CH_3F, from unity at low pressures to nearly one hundred at high pressures (Fig.6.14). The authors interpret their results in terms of swept-gain superradiance [6.50]. In a cell of 6-m length the transition occurs at a pressure p = 0.2 torr. The molecular transition is pressure broadened, with $pT_1 = pT_2 = 8$ ns torr.

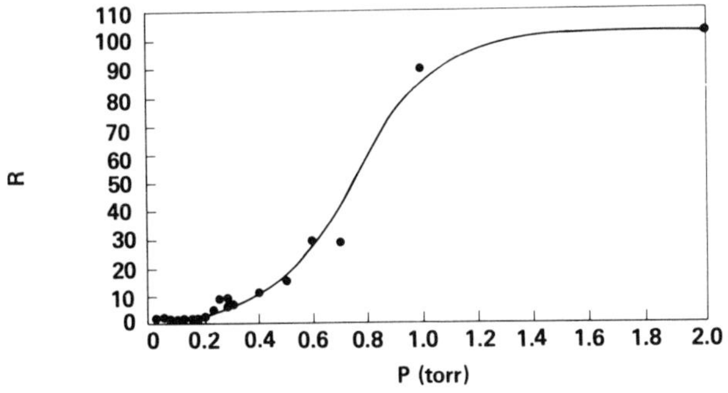

Fig.6.14. Ratio R of the measured forward-to-backward SF intensity as a function of gas pressure. The superfluorescing gas is CH_3F and the cell length 6 m. The ratio starts to deviate from 1 at p = 0.2 torr [6.25]

ROSENBERGER et al. [6.51] have derived the criterion for the onset of swept-gain superradiance to be $T_2 \simeq 2L/c$, in good agreement with the data. The peak intensity of the forward pulse is proportional to p^2 for pressures both below and above the transition point, but the proportionality constant is larger for p > 0.2 torr. The peak intensity of the backward pulse is proportional to p^2 for p < 0.2 torr and remains approximately constant for p > 0.2 torr. For p > 0.2 torr the backward pulse is longer than the forward pulse. The transition point occurs at lower pressures the longer the cell.

6.5.2 Fresnel Number and Spatial Coherence

It follows from diffraction theory that a circular disk radiating with uniform amplitude and phase emits approximately 63% of its power into a diffraction solid angle $\Omega_D = A/\lambda^2$, where A is the cross-sectional area. The dimensions of an SF sample define a geometrical solid angle $\Omega_G = A/L^2$. For Fresnel number $F \equiv A/\lambda L = 1$ the two solid angles are equal. The choice of F = 1 in an experiment is believed to optimize the cooperative emission of the sample while minimizing diffraction losses.

For F = 1 some dimensional measurements have been made by one of us [6.52]. A sample of length L = 50 mm was chosen and the pump beam diameter was set to 430 μm at half intensity to define F = 1 for the cesium wavelength of 2.931 μm. The experiment showed that 63% of the SF power was radiated within a solid angle $\Omega_{exp} = 2.8\ \lambda/L$. The actual emitting area A_{exp} could not be measured as accurately, $A_{exp} \approx (0.5 - 1)\lambda L$. It follows that $\Omega_{exp} \cdot A_{exp} \approx (1.4 - 2.8)\lambda^2$, imlying that for F = 1 the emission occurs in one or at most a few modes of the field. It has been suggested by ERNST et al. [6.53] that the emission will still occur in a single mode even with $F \gg 1$.

According to these authors a "ray" will be formed with cross-sectional area A equal to that of the excited cylinder and with divergence solid angle $\Omega_D = \lambda^2/A$. Since now $\Omega_D \ll \Omega_G$, the direction in which the "ray" is fired will fluctuate from shot to shot. In contrast, BONIFACIO [6.54] has predicted that emission will take place in F^2 modes, filling the full geometrical solid angle in each shot. The controversy still existed at the end of 1976 [6.55]. The issue has now been settled [6.37]. In the cesium experiment a cylinder was excited with L = 50 mm and $F \approx 15$. Both the emission angle and the diameter of a one-to-one image of the SF source were measured at one half of maximum intensity, with the result $\Omega_{exp} = 50\lambda/L$ and $A_{exp} = 8\lambda L$ or $\Omega_{exp} \cdot A_{exp} = 400\lambda^2$. The number of modes, 400, is indeed of the order of $F^2 = 225$.

To further test the ray concept one-half of the solid angle was focused on a first detector, and the other half on a second detector (Fig.6.15). If the ray exists, then in most shots all the energy should fall on one of the detectors, in other words, the signals should be anticorrelated. In fact the signals proved to have a strong positive correlation (+0.85) which implies that the full solid angle is filled in each individual shot.

No multiple pulsing was observed for larger F; apparently the competition of many modes leads to one single smooth pulse [6.37]. The number of atoms in the sample increases linearly with F at constant density. The spontaneous emission which initiates the SF increases as the number of modes, i.e., as F^2. Thus for large F the initiation is relatively stronger. Indeed, for F = 15 the delay times were found to be 1.5 to 2 times smaller than for F = 1 at the same density and sample length [6.37].

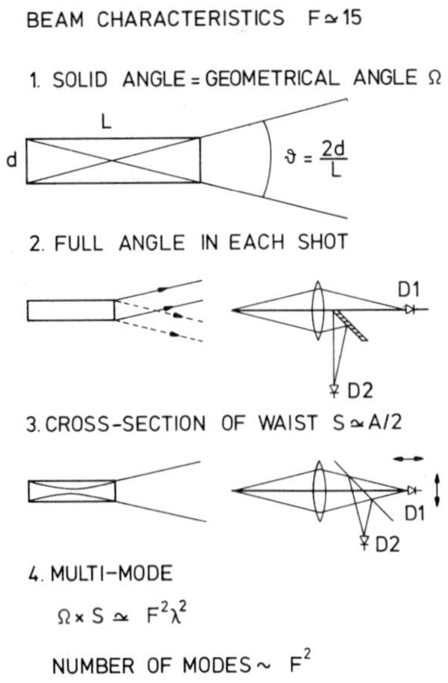

Fig.6.15. Observed SF radiation characteristics for Fresnel number $F \approx 15$. (1) The output solid angle corresponds roughly to the geometrical angle indicated. (2) A collimated "ray", switching direction from shot to shot, is not observed. Instead the full angle is filled in each shot. (3) The area S of the image of the SF source corresponds to about one-half the cross-sectional area of the pumped volume. (4) The number of modes is of the order of F^2 [6.52]

In conclusion, for $F = 1$ the radiation is emitted in one or at most a few modes and may be said to have a fair degree of spatial coherence. For $F > 1$ the radiation is emitted in F^2 modes and spatial coherence is lost. It is clear from these results that a one-dimensional description of SF is of necessity a rather crude approximation. Transverse effects are important both for small and large F.

6.6 Level Degeneracies

For a study of the basic SF phenomenon a system with nondegenerate upper and lower levels should be used. In the cesium experiment care has been taken to prepare such a system. However, interesting phenomena can be observed when the levels are degenerate or when there are several, closely spaced, upper and/or lower levels. In the following sections three classes of phenomena are considered: quantum beats on coupled transitions, beats on independent transitions, and polarization effects.

6.6.1 Quantum Beats from Coupled Transitions

Quantum beats are well known in beam foil and laser spectroscopy. In single-atom spontaneous emission [6.56], such beats can be observed only for V-type three-level systems, i.e., systems with two nearly degenerate upper levels a and b both radiatively coupled to a common lower level c. A coherent superposition of a and b must have been excited initially. The beats can be understood as arising from a modulation of the dipole moment between the mixed initial state and the final state. In superradiant and SF emission, beats can be observed not only for V-type systems, but also for Λ-type systems (one upper level and two nearly degenerate lower levels) and even for uncoupled, independent transitions. The various beats have been studied in some detail for photon echoes [6.57]. In this section SF quantum beats in a V-type system are described. No SF beats in Λ-type systems seem to have been reported to date. Beats on independent transitions are discussed in the next section.

Quantum beats in SF were first observed in cesium by VREHEN et al. [6.20]. The HFS of the relevant energy levels is shown in Fig.6.16. With the short pump pulse (1.5-2 ns) coherent superpositions could be prepared of the four sublevels of $7\ ^2P_{3/2}$ or of the two sublevels of $7\ ^2P_{1/2}$, starting from either of the ground-state sublevels. SF emission from $7\ ^2P_{1/2}$ shows modulation at the upper level difference frequency of 401 MHz, as can be seen very clearly in Fig.6.17. The single-atom quantum beat fluorescence is shown in the inset. Constructive interference is predicted at t = 0 and this should hold for SF too. The timing of the SF experiment is not sufficiently accurate to test that prediction. Possible modulation at the lower level difference frequency of 2175 MHz could not be detected in the experiment because of the finite bandwidth (~300 MHz) of the detection system. The SF from $7\ ^2P_{3/2}$ is more complicated because three F states are excited simultaneously, yielding 3 upper state beat frequencies [6.58] (Fig.6.16). Interferences between these frequencies can lead to complicated decay curves even in the single-atom case. Figure 6.18 shows both the calculated decay curves I_F for single-atom spontaneous emission under weak excitation conditions and the SF beats observed experimentally. The SF curves are quite different from the I_F curves, partly because strong excitation conditions prevail, but mainly because of the highly nonlinear process of SF evolution. Still, peaks in the SF coincide with constructive interference peaks in I_F.

6.6.2 Beats from Independent Transitions

In their work on SF in cesium the present authors also discovered beats from independent transitions [6.20]. Two upper levels a and b were excited incoherently and decayed to different final levels, c and d respectively. Beats were observed at the difference frequency $\omega_{ac} - \omega_{bd}$. They can be understood most simply as the interference between two temporally coherent waves which beat in the detector just as might be the case with the interference between two laser beams. The ensemble of atoms excited incoherently to the levels a and b can be looked upon as a mixture in which some atoms are in the state a and others are in the state b. Beats between different groups of atoms were demonstrated even more clearly in an ingenious SF experiment by GROSS et al. [6.28], also in cesium. A 4-ns-duration narrow-band pump laser pulse was tuned midway between the F=3 and F=4 levels of $7\ ^2P_{1/2}$ from either one of the ground-state hyperfine levels. The Doppler width of the pump transition is about 500 MHz in a cell. Thus the longitudinal pump pulse excites two velocity packets. A packet with forward velocity (relative to the pump pulse propagation

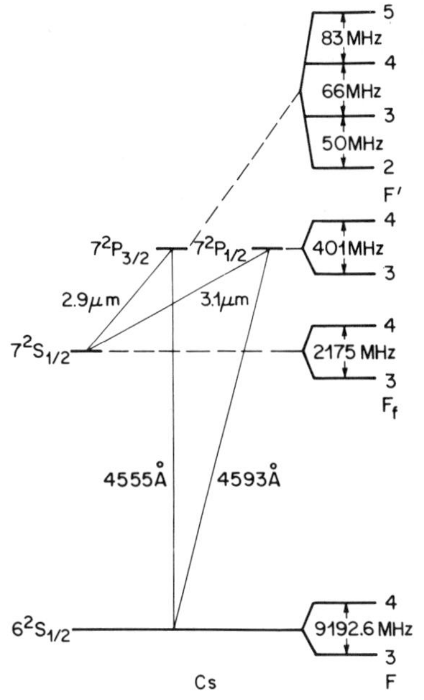

Fig.6.16. Hyperfine structure of the cesium levels relevant to quantum beat SF

Fig.6.17. Modulation of the SF emission on the $7\ ^2P_{1/2}$ to $7\ ^2S_{1/2}$ transition in cesium. The beat frequency corresponds to the 401 MHz hyperfine splitting of the upper level. Excitation with circular polarization from the F=3 ground-state level. The inset shows I_F, the single-atom quantum beat fluorescence [6.20]

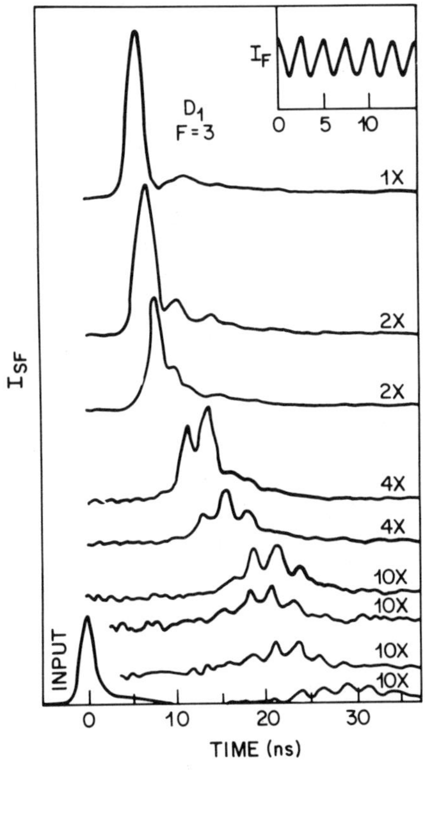

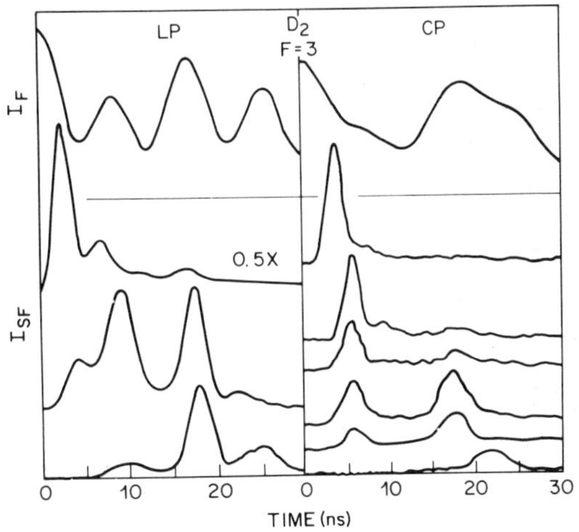

Fig.6.18. Quantum beats in the SF emission on the $7^2P_{3/2}$ to $7\ ^2S_{1/2}$ transition in cesium. Excitation from the F = 3 ground-state level with linearly polarized (LP) light on a 10-cm cell and circularly polarized (CP) light on a 2.6-cm-pathlength atomic beam. The top (I_F) curves show calculated single-atom beats. The lower curves are the observed SF beats. SF is strongly suppressed near the minima in I_F. In the CP case the delay as a function of density jumps discontinuously, avoiding the minimum at 12 ns [6.20]

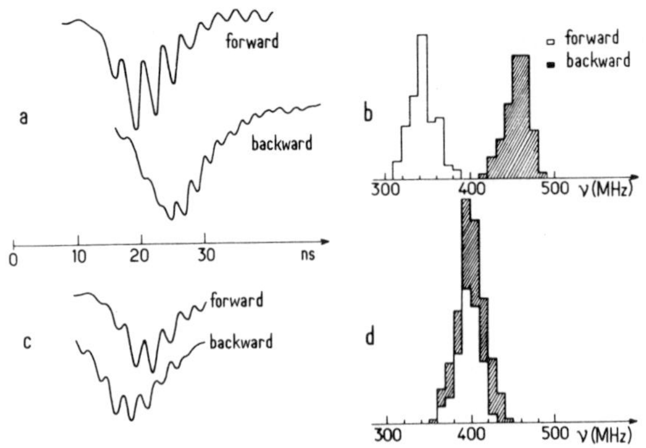

Fig.6.19a-d. Beats between different groups of Cs atoms. (a) Doppler-shifted beats in forward and backward directions following excitation with a narrow-band pulse. (b) Histogram of beat frequencies in case of narrow-band pumping. White and hatched boxes represent forward and backward signals, respectively. (c) and (d) as (a) and (b) but for broadband excitation [6.28]

direction) is excited to the lower level F = 3 and a packet with backward velocity is excited to the higher level F = 4. Both packets emit SF in both forward and backward directions. The waves in the forward direction beat together and so do the waves in the backward direction. However, because of different Doppler shifts the beat frequencies are different in the two directions (Fig.6.19). Recently MAREK [6.32] has reported evidence for beats in SF between different isotopes in rubidium.

For spectroscopic applications beats in cooperative emission have both advantages and disadvantages as compared to beats in single-atom spontaneous emission. The cooperative emission greatly enhances the intensity both by time and angle compression, and thus transitions can be studied that are too weak in spontaneous emission; this is particularly true in the infrared. Thus, SF beats might be useful for the investigation of some highly forbidden transitions of astrophysical interest for which the moderate resolution (see below) would still be adequate.

Furthermore, lower level splittings can be studied and even frequency differences between physically different atoms. On the other hand it must be noted that the spectral resolution of SF beats is limited. The beat frequency must be larger than the inverse SF pulse duration, which itself cannot be much longer than the Doppler dephasing time, and is certainly much shorter than the natural lifetime of the upper level. Moreover, in SF the atomic evolution takes place in the presence of a strong collective driving field which may affect the beat frequency (chirps). Such possible frequency shifts have not been studied in detail yet. The work of GROSS et al. [6.28] serves as a warning that Doppler shifts may also be significant, at least in a vapor.

The beats from different atomic species give support to the concept that in SF a classical coherent field is emitted. They do not constitute a definite proof, however, because all atoms are contained in the same volume and subject to the common field. Identical rapid phase fluctuations on both frequency components would destroy the temporal coherence while retaining the beats. Recently, one of the present authors [6.49] has demonstrated the existence of temporal coherence by

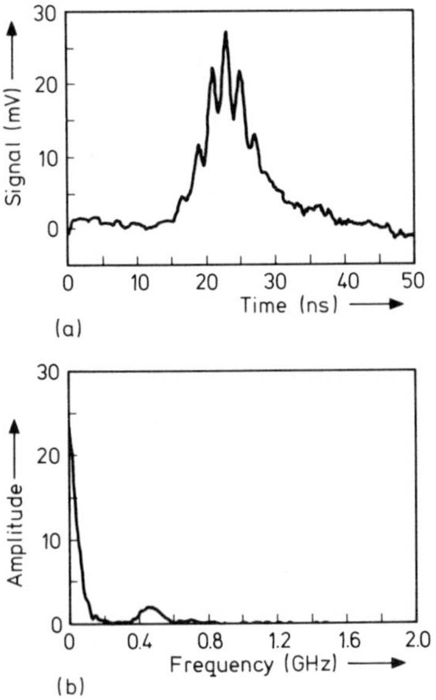

Fig.6.20a,b. Beats observed in the superposition of two SF beams, generated in two different cesium cells, magnetically tuned to different transitions [6.49]

beating together the SF outputs from two different cesium cells, magnetically tuned to different transitions with slightly different frequencies. An example of such beats is shown in Fig.6.20. Any rapid phase fluctuations that might exist would certainly be different for the two cells and thus wipe out the beats. The observation of the beats is thus proof of the temporal coherence.

6.6.3 Polarization Effects

Beats in SF arise in the presence of nearly degenerate levels. The consequences of full level degeneracies have been studied theoretically by CRUBELLIER [6.59] and by CRUBELLIER and SCHWEIGHOFER [6.60]. Those studies have been made in the small sample ($<\lambda^3$) approximation. They provide only a qualitative insight. Several aspects of the theory have been verified experimentally by CRUBELLIER et al. [6.29] on the polarization effects in SF on the transitions $6\ P_{3/2} \rightarrow 6\ S_{1/2}$, $6\ P_{1/2} \rightarrow 6\ S_{1/2}$ and $6\ S_{1/2} \rightarrow 5\ P_{3/2}$ in rubidium.

The most important features can be understood from fairly simple arguments. Let us define a coordinate system (x,y,z) with z along the sample axis. Cooperative emission will be polarized in the xy plane. To the extent that SF depopulates the upper state, z-polarized emission (single-atom spontaneous emission) will always be quenched. Now consider a coupled transition of Λ type, in which atoms may decay from an upper level a to two degenerate lower levels c and d, with orthogonal polarizations. If now $|p_{ac}| > |p_{ad}|$, then SF will evolve more rapidly on the a-c transition and as a consequence the a-d transition will be quenched. Thus the polarization will be determined completely by the a-c transition. A more subtle form of quenching occurs, e.g., when level a is coupled to c with x polarization while level b is coupled to d with x polarization and to e with y polarization. If now a decays rapidly by SF, a strong x-polarized electric field will be set up in the sample and this field may induce a preferential decay from b to d even when $|p_{be}| > |p_{bd}|$.

An interesting case arises in the decay of two uncoupled transitions a → c and b → d with orthogonal polarizations say along x and y, respectively. If $|p_{ac}| = |p_{bd}|$ and if the initial populations of the two upper levels are the same, then spontaneous emission will be completely unpolarized. In SF, however, coherent fields develop on both transitions and the emission will in general be elliptically polarized. From shot to shot the phase will vary randomly and so will the polarization of the output.

Finally a coupled transition of V type may be studied in which two degenerate upper levels a and b are both coupled to a common lower level c. If in that case a coherent superposition is excited, the emission will be fully polarized and the polarization will be determined by the mixing coefficients of the initial state.

6.7 Summary

The recent experiments on SF have clarified many details of this interesting phenomenon. A temporally coherent wave is emitted with a fair degree of spatial coherence (for Fresnel number 1). The initiation seems to be well described by the one-dimensional fully retarded quantum-mechanical theories now available. The effective initial tipping angle is of order $2/\sqrt{N}$ and the standard deviation of the delay time due to quantum fluctuations is of the order of 12% for cesium ($N \approx 10^8$). In the nonlinear stage of the pulse evolution the transverse effects play an important part. These transverse effects are particularly manifest in the multiple pulsing which occurs for $\tau_R < 2L/c$. The theory of transverse effects is still in its infancy. Present day theories cannot predict the observed pulse shapes satisfactorily.

For $F \gg 1$ the samples emit in many modes, the spatial coherence is lost and the delay time reduced. For very long samples swept-gain superradiance can be observed. In the presence of nearly degenerate levels, beats can be seen both on coupled transitions and on independent transitions, which in principle allow the measurement of isotope shifts.

It is clear that considerable progress has already been made, but further interesting developments may be anticipated. Transverse effects will be included in the theory and the results may suggest new experiments. The study of fluctuations has only just begun. The advent of visible SF will allow the use of sensitive photon counting methods and photographic techniques for the study of photon statistics and intereference patterns. The precise role of inhomogeneous line broadening may be pinned down, and with that the transition from SF to amplified spontaneous emission (ASE) may become better understood. Evolution of the SF emission during the pump pulse (Raman type SF), transverse excitation of very long samples, cascade SF —all these phenomena deserve further investigation. And finally SF still awaits application, in spectroscopy or for the generation of very short pulses.

References

6.1 R. Bonifacio, L.A. Lugiato: Phys. Rev. A*11*, 1507 (1975); *12*, 587 (1975)
6.2 R.H. Dicke: Phys. Rev. *93*, 99 (1954)
6.3 R.H. Dicke: In *Quantum Electronics*, ed. by P. Grivet, N. Bloembergen (Dunod, Paris 1963) Vol.1, p.35
6.4 D. Polder, M.F.H. Schuurmans, Q.H.F. Vrehen: Phys. Rev. A*19*, 1192 (1979)
6.5 R. Friedberg, S.R. Hartmann: Opt. Commun. *10*, 298 (1974)
6.6 J.C. MacGillivray, M.S. Feld: Phys. Rev. A*14*, 1169 (1976)
6.7 R. Glauber, F. Haake: Phys. Lett. *68*A, 29 (1978)
6.8 M.F.H. Schuurmans, D. Polder, Q.H.F. Vrehen: J. Opt. Soc. Am. *68*, 699 (1978)
6.9 N. Skribanowitz, I.P. Herman, J.C. MacGillivray, M.S. Feld: Phys. Rev. Lett. *30*, 309 (1973)
6.10 I.P. Herman, J.C. MacGillivray, N. Skribanowitz, M.S. Feld: In *Laser Spectroscopy*, ed. by R.G. Brewer, A. Mooradian (Plenum, New York 1974)
6.11 H.M. Gibbs, Q.H.F. Vrehen, H.M.J. Hikspoors: Phys. Rev. Lett. *39*, 547 (1977)
6.12 Q.H.F. Vrehen, M.F.H. Schuurmans: Phys. Rev. Lett. *42*, 224 (1979)
6.13 F.T. Arecchi, E. Courtens: Phys. Rev. A*2*, 1730 (1970)
6.14 N.E. Rehler, J.H. Eberly: Phys. Rev. A*3*, 1735 (1971); J.H. Eberly: Am. J. Phys. *40*, 1374 (1972)
6.15 M. Gross, C. Fabre, P. Pillet, S. Haroche: Phys. Rev. Lett. *36*, 1035 (1976)
6.16 A. Flusberg, T. Mossberg, S.R. Hartmann: Phys. Lett. *58*A, 373 (1976)
6.17 T.W. Karras, R.S. Anderson, B.G. Bricks, C.E. Anderson: In Ref.6.18, p.101
6.18 C.M. Bowden, D.W. Howgate, H.R. Robl (eds.): *Cooperative Effects in Matter and Radiation* (Plenum, New York 1977)
6.19 C.M. Bowden, C.C. Sung: Phys. Rev. A*18*, 1558 (1978)
6.20 Q.H.F. Vrehen, H.M.J. Hikspoors, H.M. Gibbs: Phys. Rev. Lett. *38*, 764 (1977)
6.21 S.L. McCall, E.L. Hahn: Phys. Rev. Lett. *18*, 908 (1967); Phys. Rev. *183*, 457 (1969)
6.22 D.C. Burnham, R.Y. Chiao: Phys. Rev. *188*, 667 (1969)
6.23 H.M. Gibbs, Q.H.F. Vrehen, H.M.J. Hikspoors: In *Laser Spectroscopy III*, ed. by J.L. Hall, J.L. Carlsten, Springer Series in Optical Science, Vol.7 (Springer, Berlin, Heidelberg, New York 1977) p.213
6.24 A.T. Rosenberger, S.J. Petuchowski, T.A. DeTemple: In Ref.6.18, p.15
6.25 J.J. Ehrlich, C.M. Bowden, D.W. Howgate, S.H. Lehnigk, A.T. Rosenberger, T.A. DeTemple: In Ref.6.26, p.923
6.26 L. Mandel, E. Wolf (eds.): *Coherence and Quantum Optics IV* (Plenum, New York 1978)
6.27 J.C. MacGillivray, M.S. Feld: Appl. Phys. Lett. *31*, 74 (1977); F.A. Hopf, P. Meystre, M.O. Scully, J.F. Seely: Phys. Rev. Lett. *35*, 511 (1975)
6.28 M. Gross, J.M. Raimond, S. Haroche: Phys. Rev. Lett. *40*, 1711 (1978)
6.29 A. Crubellier, S. Liberman, P. Pillet: Phys. Rev. Lett. *41*, 1237 (1978)
6.30 J. Okada, K. Ikeda, M. Matsuoka: Opt. Commun. *26*, 189 (1978); *27*, 321 (1978)
6.31 M.F.H. Schuurmans, D. Polder: Phys. Lett. *72*A, 306 (1979)
6.32 J. Marek: J. Phys. B*12*, L229 (1979)
6.33 C. Brechignac, Ph. Cahuzac: J. Phys. Paris Lett. *40*, L-123 (1979)
6.34 Ph. Cahuzac, H. Sontag, P.E. Toschek: Opt. Commun. *31*, 37 (1979)
6.35 H.M. Gibbs: In Ref.6.18, p.61

6.36 Q.H.F. Vrehen: In Ref.6.18, p.79
6.37 Q.H.F. Vrehen, H.M.J. Hikspoors, H.M. Gibbs: In Ref.6.26, p.543
6.38 H.M. Gibbs: In *Coherence in Spectroscopy and Modern Physics*, ed. by F.T. Arecchi, R. Bonifacio, M.O. Scully (Plenum, New York 1977) p.121
6.39 Q.H.F. Vrehen, H.M. Gibbs: J. Opt. Soc. Am. *68*, 699 (1978)
6.40 Q.H.F. Vrehen: In *Trends in Physics, 1978*, ed. by M.M. Woolfson (Adam Hilger, Bristol 1979) p.95
6.41 J.P. Wittke, P.J. Warter: J. Appl. Phys. *35*, 1668 (1964);
F.T. Arecchi, R. Bonifacio: IEEE J. Quantum Electron. *1*, 169 (1965);
F.A. Hopf, M.O. Scully: Phys. Rev. *179*, 399 (1969);
A. Icsevgi, W.E. Lamb, Jr.: Phys. Rev. *185*, 517 (1969)
6.42 S.L. McCall: Dissertation, University of California (1968)
6.43 R. Saunders, S.S. Hassan, R.K. Bullough: J. Phys. A*9*, 1725 (1976);
R. Saunders, R.K. Bullough: In Ref.6.18, p.209;
R.K. Bullough, R. Saunders, C. Feuillade: In Ref.6.26, p.263
6.44 H.M. Gibbs, B. Bölger, F.P. Mattar, M.C. Newstein, G. Forster, P.E. Toschek: Phys. Rev. Lett. *37*, 1743 (1976);
N. Wright, M.C. Newstein: Opt. Commun. *9*, 8 (1973);
F.P. Mattar, M.C. Newstein: Opt. Commun. *18*, 70 (1976); In Ref.6.18, p.139;
F.P. Mattar, M.C. Newstein, P.E. Serafim, H.M. Gibbs, B. Bölger, G. Forster, P.E. Toschek: In Ref.6.26, p.143
6.45 M.F.H. Schuurmans, D. Polder: In *Laser Spectroscopy IV*, ed. by H. Walther, K.W. Rothe, Springer Series in Optical Sciences, Vol.21 (Springer, Berlin, Heidelberg, New York 1979)
6.46 F. Haake, H. King, G. Schröder, J. Haus, R. Glauber, F. Hopf: Phys. Rev. Lett. *42*, 1740 (1979)
6.47 V. Degiorgio: Opt. Commun. *2*, 362 (1971)
6.48 B. Bölger, L. Baede, H.M. Gibbs: Opt. Commun. *18*, 67 (1976)
6.49 Q.H.F. Vrehen: In *Laser Spectroscopy IV*, ed. by H. Walther, K.W. Rothe, Springer Series in Optical Sciences, Vol.21 (Springer, Berlin, Heidelberg, New York 1979)
6.50 R. Bonifacio, F.A. Hopf, P. Meystre, M.O. Scully: Phys. Rev. A*12*, 2568 (1975);
F.A. Hopf, P. Meystre: Phys. Rev. A*12*, 2534 (1975);
F.A. Hopf, P. Meystre, D.W. McLaughlin: Phys. Rev. A*13*, 777 (1976)
6.51 A.T. Rosenberger, T.A. DeTemple, C.M. Bowden, C.C. Sung: J. Opt. Soc. Am. *68*, 700 (1978)
6.52 Q.H.F. Vrehen: Unpublished results
6.53 V. Ernst, P. Stehle: Phys. Rev. *176*, 1456 (1968);
V. Ernst: Z. Phys. *229*, 432 (1969)
6.54 R. Bonifacio: In *Cooperative Effects, Progress in Synergetics*, ed. by H. Haken (North Holland, Amsterdam 1974) pp.97-117, esp. p.112
6.55 Ref.6.18, p.380 (Discussion)
6.56 S. Haroche: In *High-Resolution Laser Spectroscopy*, ed. by K. Shimoda, Topics in Applied Physics, Vol.13 (Springer, Berlin, Heidelberg, New York 1976)
6.57 L.Q. Lambert, A. Compaan, I.D. Abella: Phys. Lett. *30*A, 153 (1969); Phys. Rev. A*4*, 2022 (1971);
P.F. Liao, P. Hu, R. Leigh, S.R. Hartmann: Phys. Rev. A*9*, 332 (1974);
R.L. Shoemaker, F.A. Hopf: Phys. Rev. Lett. *33*, 1527 (1974);
I.D. Abella, A. Compaan, L.Q. Lambert: In *Laser Spectroscopy*, ed. by R.G. Brewer, A. Mooradian (Plenum, New York 1974);
T. Baer, I.D. Abella: Phys. Lett. *59*A, 371 (1976)
6.58 S. Haroche, J. Paisner, A.L. Schawlow: Phys. Rev. Lett. *30*, 948 (1973)
6.59 A. Crubellier: Phys. Rev. A*15*, 2430 (1977)
6.60 A. Crubellier, M.G. Schweighofer: Phys. Rev. A*18*, 1797 (1978)

Additional References with Titles

T. Baba, K. Ikeda: Fluctuation of polarized light in cooperative spontaneous emission. J. Phys. Soc. Jpn. *50*, 217 (1981)

R. Bonifacio, J.D. Farina, L.M. Narducci: Transverse effects in superfluorescence. Opt. Commun. *31*, 377 (1979)

C. Brechignac, Ph. Cahuzac: Population inversion on the resonance line of strontium by using cascading superfluorescence in a three-level system. J. Phys. B*14*, 221 (1981)

N.W. Carlson, D.J. Jackson, A.L. Schawlow, M. Gross, S. Haroche: Superradiance triggering spectroscopy. Opt. Commun. *32*, 350 (1980)

A. Crubellier, C. Brechignac, P. Cahuzac, P. Pillet: "Coupled transitions in Superradiance", in *Laser Spectroscopy IV*, ed. by H. Walther, K.W. Rothe, Springer Series in Optical Sciences, Vol.21 (Springer, Berlin, Heidelberg, New York 1979)

A. Crubellier, S. Liberman, P. Pillet, M.G. Schweighofer: Experimental study of quantum fluctuations of polarisation in superradiance. J. Phys. B*14*, L177 (1981)

G. Dodel, G. Magyar, D. Veron: Oscillator and superradiance characteristics of a "zig-zag" pumped 66-nm D_2O-laser. Infrared Phys. *18*, 529 (1978)

F. Gounand, M. Hugon, P.R. Fournier, J. Berlande: Superradiant cascading effects in rubidium Rydberg levels. J. Phys. B*12*, 547 (1979)

M. Gross, P. Goy, C. Fabre, S. Haroche, J.M. Raimond: Maser oscillation and microwave superradiance in small systems of Rydberg atoms. Phys. Rev. Lett. *43*, 343 (1979)

F. Haake, H. King, G. Schröder, J. Haus, R. Glauber: Fluctuations in superfluorescence. Phys. Rev. A*20*, 2047 (1979)

F. Haake, J. Haus, H. King, G. Schröder, R. Glauber: Delay time statistics and inhomogeneous line broadening in superfluorescence. Phys. Rev. Lett. *45*, 558 (1980)

F. Haake, J.W. Haus, H. King, G. Schröder, R. Glauber: Delay time statistics of superfluorescent pulses. Phys. Rev. A*23*, 1322 (1981)

J.A. Hermann: An amplifying solution of the Maxwell-Bloch equations with atomic relaxation and field losses. J. Phys. A*13*, 3543 (1980)

K. Ikeda, J. Okada, M. Matsuoka: Theory of cooperative cascade emission. I. Linear stochastic theory. J. Phys. Soc. Jpn. *48*, 1636 (1980)

K. Ikeda, J. Okada, M. Matsuoka: Theory of cooperative cascade emission. II. Nonlinear evolution. J. Phys. Soc. Jpn. *48*, 1646 (1980)

J.C. MacGillivray, M.S. Feld: Limits of superradiance as a process for achieving short pulses of high energy. Phys. Rev. A*23*, 1334 (1981)

J.C. MacGillivray, M.S. Feld: Superradiance in atoms and molecules. Contemp. Phys. *22*, 299 (1981)

J.C. MacGillivray, M.S. Feld: "Superradiance", in *Coherent Nonlinear Optics. Recent Advances*, ed. by M.S. Feld, V.S. Lethokov, Topics in Current Physics, Vol.21 (Springer, Berlin, Heidelberg, New York 1980)

J. Marek, M. Ryschka: Quantum beats in superradiance in sodium vapours. J. Phys. B*13*, L491 (1980)

F.P. Mattar, H.M. Gibbs, S.L. McCall, M.S. Feld: Transverse effects in superfluorescence. Phys. Rev. Lett. *46*, 1123 (1981)

M.F.H. Schuurmans: Superfluorescence and amplified spontaneous emission in an inhomogeneously broadened medium. Opt. Commun. *34*, 185 (1980)

Subject Index

Springer Series in Synergetics

Series Editor: H. Haken

Volume 1
H. Haken
Synergetics
An Introduction
Nonequilibrium Phase Transitions and Self-Organization in Physics, Chemistry and Biology
2nd enlarged edition. 1978. 152 figures, 4 tables. XII, 355 pages
ISBN 3-540-08866-0

Volume 2
Synergetics
A Workshop
Proceedings of the International Workshop on Synergetics at Schloss Elmau, Bavaria, May 2–7, 1977
Editor: H. Haken
1977. 136 figures. VIII, 274 pages
ISBN 3-540-08483-5

Volume 3
Synergetics Far from Equilibrium
Proceedings of the Conference Far from Equilibrium: Instabilities and Structures, Bordeaux, France, September 27–29, 1978
Editors: A. Pacault, C. Vidal
1979. 109 figures, 3 tables. IX, 175 pages
ISBN 3-540-09304-4

Volume 4
Structural Stability in Physics
Proceedings of Two International Symposia on Applications of Catastrophe Theory and Topological Concepts in Physics, Tübingen, Federal Republic of Germany, May 2–6 and December 11–14, 1978
Editors: W. Güttinger, H. Eikemeier
1979. 108 figures, 8 tables. VIII, 311 pages
ISBN 3-540-09463-6

Volume 5
Pattern Formation by Dynamic Systems and Pattern Recognition
Proceedings of the International Symposium on Synergetics at Schloß Elmau, Bavaria, April 30 – May 5, 1979
Editor: H. Haken
1979. 156 figures, 16 tables. VIII, 305 pages
ISBN 3-540-09770-8

Volume 6
Dynamics of Synergetic Systems
Proceedings of the International Symposium on Synergetics, Bielefeld, Federal Republic of Germany, September 24–29, 1979
Editor: H. Haken
1980. 146 figures, some in color, 5 tables. VIII, 271 pages
ISBN 3-540-09918-2

Volume 7
L. A. Blumenfeld
Problems of Biological Physics
1981. 38 figures. IX, 224 pages
ISBN 3-540-10401-1

Volume 8
Stochastic Nonlinear Systems
in Physics, Chemistry, and Biology
Proceedings of the Workshop, Bielefeld, Federal Republic of Germany, October 5–11, 1980
Editors: L. Arnold, R. Lefever
1981. 48 figures. VIII, 237 pages
ISBN 3-540-10713-4

Volume 9
Numerical Methods in the Study of Critical Phenomena
Proceedings of a Colloquium, Carry-le-Rouet, France, June 1980
Editors: J. Della Dora, J. Demongeot, B. Lacolle
1981. 83 figures. IX, 269 pages
ISBN 3-540-11009-7

Volume 10
Y. L. Klimontovich
The Kinetic Theory of Electromagnetic Processes
In preparation

Volume 11
Chaos and Order in Nature
Proceedings of the International Symposium on Synergetics at Schloss Elmau, Bavaria, April 27–May 2, 1981
Editor: H. Haken
1981. 134 figures. VIII, 275 pages
ISBN 3-540-11101-8

Springer-Verlag
Berlin
Heidelberg
New York